SpringerBriefs in Molecular Science

Biobased Polymers

Series editor

Patrick Navard, Sophia Antipolis cedex, France

Published under the auspices of EPNOE*Springerbriefs in Biobased polymers covers all aspects of biobased polymer science, from the basis of this field starting from the living species in which they are synthetized (such as genetics, agronomy, plant biology) to the many applications they are used in (such as food, feed, engineering, construction, health, …) through to isolation and characterization, biosynthesis, biodegradation, chemical modifications, physical, chemical, mechanical and structural characterizations or biomimetic applications. All biobased polymers in all application sectors are welcome, either those produced in living species (like polysaccharides, proteins, lignin, …) or those that are rebuilt by chemists as in the case of many bioplastics.

Under the editorship of Patrick Navard and a panel of experts, the series will include contributions from many of the world's most authoritative biobased polymer scientists and professionals. Readers will gain an understanding of how given biobased polymers are made and what they can be used for. They will also be able to widen their knowledge and find new opportunities due to the multidisciplinary contributions.

This series is aimed at advanced undergraduates, academic and industrial researchers and professionals studying or using biobased polymers. Each brief will bear a general introduction enabling any reader to understand its topic.

EPNOE The European Polysaccharide Network of Excellence (www.epnoe.eu) is a research and education network connecting academic, research institutions and companies focusing on polysaccharides and polysaccharide-related research and business.

More information about this series at http://www.springer.com/series/15056

It-Meng Low · Ahmad Hakamy
Faiz Shaikh

High Performance Natural Fiber-Nanoclay Reinforced Cement Nanocomposites

 Springer

It-Meng Low
Department of Applied Physics
Curtin University
Perth, WA
Australia

Faiz Shaikh
Department of Civil Engineering
Curtin University
Perth, WA
Australia

Ahmad Hakamy
Department of Applied Physics
Curtin University
Perth, WA
Australia

and

Department of Physics
Umm Al-Qura University
Makkah
Saudi Arabia

ISSN 2191-5407   ISSN 2191-5415   (electronic)
SpringerBriefs in Molecular Science
ISSN 2510-3407   ISSN 2510-3415   (electronic)
Biobased Polymers
ISBN 978-3-319-56587-3   ISBN 978-3-319-56588-0   (eBook)
DOI 10.1007/978-3-319-56588-0

Library of Congress Control Number: 2017936703

Printed on acid-free paper

This Springer imprint is published by Springer Nature
The registered company is Springer International Publishing AG
The registered company address is: Gewerbestrasse 11, 6330 Cham, Switzerland

Contents

Chapter 1
Introduction

Concrete is one of the most widely used construction materials in the world. Portland cement is widely utilised in construction materials because it is the most inexpensive binder [1]. The concrete, mortar and cement paste have high compressive strength but they are characterized by low tensile strength, flexural strength and flexural and fracture toughness (much weaker in tension than in compression). The brittle concrete, mortar and cement paste matrix can be reinforced by using steel fibres, synthetic fibres (i.e. carbon or glass fibres) or natural fibres in order to produce fibre-reinforced cement composites (FRC) [2, 3].

Recently, natural fibres are gaining increasing popularity to develop 'environmental-friendly construction materials' as alternative to synthetic fibres in fibre-reinforced concrete [4, 5]. These fibres are cheaper, biodegradable and lighter than their synthetic counterparts. Some examples of natural fibres are: Sisal, Flax, Hemp, Bamboo, Coir and others [6–8]. Natural and cellulose fibres are used in polymer and cement matrices to improve their tensile and flexural strength and fracture resistance properties [5, 9, 10]. Hemp fibres are used in a number of products such as paper, textiles and ropes and recently they have been gaining increasing social acceptance of natural materials [11]. Hemp fibre has high specific tensile strength and specific modulus of elasticity that can candidate it as reinforcing fibre in fibre reinforced composites [12, 13]. Several studies have shown that hemp fibres as reinforcement in concrete improve the tensile strength, flexural strength and toughness of hemp fibres-reinforced concrete [14, 15].

On the other hand, one of the most effective techniques to obtain a high performance cementitious composite is by reinforcement with textile (fabrics), which are impregnated with cement paste or mortar. Synthetic fabrics such as polyethylene (PE) and polypropylene (PP) have been used as reinforcement for cement composites [16, 17]. This system has superior filament-matrix bonding which improve mechanical properties such as tensile and flexural strength better than continuous or short fibres [18, 19]. In contrast, the use of natural fibre sheets and fabrics is more prevalent in polymer matrix when compared to cement-based

© The Author(s) 2017
I.-M. Low et al., *High Performance Natural Fiber-Nanoclay Reinforced Cement Nanocomposites*, Biobased Polymers, DOI 10.1007/978-3-319-56588-0_1

matrix. For example, using cellulose-fibre sheets in epoxy or viny-ester matrix have resulted in significantly improved fracture toughness [20, 21].

However, the long term durability of natural fibres in cement composites has been the major issue in which it has limited their applications in cementitious composites. These issues could be the degradation of fibres in a high alkaline environment of cement composites and relatively weak interfacial bonding between the natural fibre and the cement matrix [8, 22, 23]. In order to improve the durability of fibre reinforced cement composites, there are two possible methods: (i) modification of fibre surfaces and (ii) modification of the cement matrix [24, 25].

Concerning the first method, some researchers have showed that pre-treatments of natural fibre surfaces via some chemical agents such as alkalization, PEI (polyethylene imine) and $Ca(OH)_2$ have slightly improved the interfacial bonding between the natural fibres and the cement matrix. As a result, the mechanical properties of such composites are enhanced [26–28]. Regarding the second approach, alkaline condition in cement can be reduced by using pozzolanic supplementary cementitious materials (SCMs) such as silica fume, fly ash, metakaolin and amorphous nanomaterials. Thus this approach could improve interfacial bond, mechanical properties and durability of natural fibre-reinforced cement composites [29, 30]. However, the extent of substitution is limited due to the reduction of early strength [31–34].

Nowadays, nanotechnology is one of the most active research areas in the civil engineering and construction materials [35, 36]. In the construction industry, several types of nanomaterials have been incorporated into concretes such as nano-SiO_2, carbon nanotubes, nano-$CaCO_3$, nano-Al_2O_3 and nano-TiO_2 in order to improve the durability and mechanical properties of concrete and Portland cement matrix [37–39]. Supit and Shaikh [40] reported that the addition of 1% nano-$CaCO_3$ improved the compressive strength of mortar and concrete significantly. Nanoclay (NC) is a new generation of processed clay for a wide range of high-performance cement nanocomposite. Some examples of nanoclay are nano-halloysite, nano-cloisite 30B and nano-kaolin. As a kind of nano-pozzolanic material, nanoclay not only reduces the pore size and porosity of the cement matrix, but also improves the strength of cement matrix through pozzolanic reactions [35, 41–44]. Calcined nanoclay is prepared by heating nanoclay at certain temperature for certain period of time in order to transform nanoclay to amorphous state in which the calcined nanoclay can behave as a highly reactive artificial pozzolan such as silica fume, metakaolin, nano-SiO_2, and nano-metakaolin [45–47].

The use of nanoclay and calcined nanoclay in natural fibre reinforced cement composite especially treated hemp fabric-reinforced cement composite can be expected to reduce the alkalinity of cement matrix by consuming calcium hydroxide during pozzolanic reaction. Hence it can improve the durability of natural fibre in the composite and the microstructure of nanoclay-cement matrix due to particle packing (filler effect) of nanoclay as well as improves its bond with natural fibre. Thus, better composite behaviour can be expected in natural fibre reinforced nanoclay-cement composite, particularly treated hemp fabric-reinforced calcined nanoclay-cement composite.

In this book, we describe a novel method for synthesizing hemp fabric-reinforced cement nanocomposites. In order to improve the durability and degradation resistance of hemp fibre in hemp fibre-reinforced cement composites, the combination of both methods (modification of hemp fibre surfaces and the cement matrix) were used. Nano-particles such as nanoclay was used to partially substitute Portland cement at various ratios. Hemp fibres were utilised to reinforce this nanoclay-cement matrix. NaOH pre-treatments of fibre surfaces and thermal pre-treatments of nanoclay were employed in order to enhance the microstructure and mechanical and durability properties. Therefore, this work evaluated the effect of different nanoclay and calcined nanoclay contents on various mechanical and physical properties of cement matrix and untreated and treated hemp fabric reinforced cement composites. The effect of nanoclay and calcined nanoclay on thermal behaviour of untreated and treated hemp fabric reinforced composites was also evaluated. The effect of calcined nanoclay on the durability properties of treated hemp fabric-reinforced cement composite was studied. The durability of the treated hemp fabric-reinforced cement composites and nanocomposites was discussed based on the porosity and flexural strength obtained at 56 days and 236 days (3 wet/dry cycles). The microstructures of nanocomposites and untreated and treated—reinforced cement nanocomposites were also investigated using High resolution transmission electron microscopy, Quantitative X-ray Diffraction Analysis with Rietveld refinement by Bruker DIFFRAC^{plus} TOPAS software, X-ray Diffraction, Synchrotron radiation diffraction, Scanning electron microscopy, Energy dispersive spectroscopy and Thermogravimetric analysis.

This work seeks to gain a better understanding of natural fibre-reinforced nanocomposites made from Hemp fibres and nanoclay. In this work, an innovative method is utilised to synthesize hemp fibre reinforced cement composite containing nanoclay with unique mechanical properties for use as building materials. This research provides new design concepts for developing cementitious composites which will have a significant impact on the future application of this technology. Hitherto, little or no research was reported on using of both chemically treated hemp fabrics and nanoclay (or calcined nanoclay) as reinforcement in cement-composites. In this work, nanoclay was utilised as partial replacement of cement at various contents to produce the nanocomposites and hemp fabrics (HF) were used as reinforcement to fabricate HF-reinforced cement nanocomposites. Hemp fibre reinforced cement based composites are sustainable alternative to their synthetic fibre based counterparts, due to their low cost, bio-degradable, local availability and extremely less energy intensive manufacturing process. However, the bond of hemp fibres with cement matrix is relatively week and high alkalinity of cement matrix adversely affects the durability of hemp fibres in the composites, which ultimately affects the mechanical and durability properties of the composites. The use of nanoclay in hemp fibre reinforced cement composite could be expected to reduce the alkalinity of cement matrix by consuming calcium hydroxide during pozzolanic reaction, hence improves the durability of hemp fibre in the composite and the dense nanoclay-cement matrix due to particle packing of nanoclay also improves its bond with hemp fibre.

In the future, this novel natural fibre-reinforced nanoclay-cement nanocomposite could be utilised in the construction industry. Moreover, it can also be widely employed in architectural structures, block walls, pavements, bridges, roads, fences, poles, pipes and footings for gates. Furthermore, it can be utilised as an alternative to steel or synthetic fibres, which replaces glass fiber reinforced concrete in some applications including concrete tiles, roofing sheets, on-ground floors, window trims, fountains, ceilings, structural laminate and sandwich panels.

Organisation of Chapters

This book consists of seven chapters where background and research aims are introduced in this chapter. In second chapter, methodology to fabricate the unreinforced nanoclay cement matrix and HF reinforced nanoclay composite, test methods to measure the physical, mechanical and durability properties of both unreinforced and HF reinforced nanoclay composite are presented. In addition the micro-scale characterisation techniques used to characterise the matrix, HF and the interface are also presented. Finally, physical and chemical properties of all materials used in this research are also provided in this chapter.

In Chap. 3 various engineering properties of nanoclay-cement nanocomposites are presented. Ordinary Portland cement (OPC) was partially substituted by nanoclay with 1, 2 and 3 wt% of OPC. The results indicated that an optimum replacement of OPC with 1 wt% nanoclay was observed through improved reduced porosity and water absorption as well as increased density, flexural strength and fracture toughness of nanoclay cement nanocomposites. The microstructural analyses indicate that the nanoclay behaves not only as a filler to improve the microstructure, but also as activator to promote the pozzolanic reaction that led to an obvious consumption of $Ca(OH)_2$ crystals and production of more amorphous calcium silicate hydrate gel. The influence of nanoclay on properties of hemp fabric-reinforced cement composites is also presented in this chapter. Two layers of hemp fabric were used and the total amount of hemp fabric in each specimen was about 2.5 wt%. For each series, three prismatic plate specimens of 300 mm × 70 mm × 10 mm were cast then after 56 days several rectangular specimens of each series with dimensions 70 mm × 20 mm × 10 mm were cut from the fully cured prismatic plate for mechanical and physical tests. The results indicated that hemp fabric-reinforced nanocomposites containing 1 wt% nanoclay had denser microstructure than others, and thus this improvement led to enhance the hemp fabric-nanomatrix adhesion. In addition, the incorporation of 1 wt% nanoclay into the hemp fabric-reinforced nanocomposites improved the thermal stability, decreased the porosity and water absorption as well as increased the density, flexural strength, fracture toughness, impact strength and Rockwell hardness as well as improved thermal stability when compared to the hemp fabric-reinforced cement composite.

In Chap. 4, the influence of nanoclay and calcined nanoclay on the mechanical and thermal properties of cement nano-composites is presented. Calcined nanoclay was prepared by heating nanoclay at 900 °C for 2 h. Ordinary Portland cement was partially replaced by nanoclay or calcined nanoclay of 1, 2 and 3 wt% OPC and sample preparation was the same as the first part. It was found that calcined nanoclay was more effective than nanoclay in terms of its high pozzolanic reaction. The optimum content of calcined nanoclay was also found to be 1 wt%. The cement nanocomposite containing 1 wt% calcined nanoclay decreased the porosity, water absorption and increased the density, compressive strength, flexural strength, fracture toughness, impact strength and Rockwell hardness as well as improved thermal stability when compared to the control cement paste and cement nanocomposite containing 1 wt% nanoclay.

In Chap. 5, the influence of NaOH-treatment on the microstructure and mechanical properties of treated hemp fabric-reinforced cement composites is presented. The surface modification of hemp fibres was treated by immersing the fabric in 1.7 M NaOH solution for 48 h at 25 °C, followed by neutralization and drying. Samples of un-treated hemp fabric-reinforced cement composites were fabricated with various contents of hemp fabric: 4.5 wt% (4 layers of fabric), 5.7 wt% (5 layers of fabric), 6.9 wt% (6 layers of fabric) and 8.1 wt% (7 layers of fabric). For the fabrication of treated hemp fabric-reinforced cement composite samples, only 6 layers of treated hemp fabric were used because they have been shown to exhibit the best mechanical performance. The results indicated that the optimum content of hemp fabric was found to be 6.9 wt% (6 hemp fabric layers). NaOH-treated hemp fabric-reinforced cement composites exhibit the highest flexural strength and fracture toughness when compared to their non-treated counterparts. The influences of calcined nanoclay (CNC) and NaOH treatment of hemp fabric on the thermal and mechanical properties of treated hemp fabric-reinforced cement nanocomposites are also presented in this chapter. The cement matrix was fabricated as the same as in the fourth part with 1, 2 and 3 wt% of calcined nanoclay. Six layers of treated hemp fabric were used to reinforce the cement nanocomposite matrix and the total amount of treated hemp fabric in each specimen was about 6.9 wt%. Results indicated that physical, mechanical and thermal properties were enhanced due to the addition of CNC into the cement matrix and the optimum content of CNC was 1 wt%. The treated hemp fabric-reinforced nanocomposites containing 1 wt% CNC exhibited the highest flexural strength, fracture toughness, impact strength and thermal stability than their counterparts and good fibre-matrix interface.

In Chap. 6, the effect of calcined nanoclay on the durability of NaOH treated hemp fabric-reinforced cement nanocomposites is investigated. The treated hemp fabric-reinforced cement composites and nanocomposites were subjected to 3 wetting and drying cycles and then tested at 56 and 236 days. The influences of CNC dispersion on the durability of these composites have been characterized in terms of porosity, flexural strength, stress-midspan deflection curves and microstructural observation of hemp surface by SEM images. Results indicated that the CNC effectively mitigated the degradation of hemp fibre. The durability and the degradation resistance of hemp fibre were enhanced due to the addition of CNC into the cement matrix and the optimum content of CNC was 1 wt%. SEM micrographs indicated that hemp fibres in treated hemp fabric reinforced cement composites undergo more degradation that in treated hemp fabric reinforced cement nanocomposites containing 1 wt% CNC. Based on overall results, the addition of CNC has great potential to improve the durability of treated hemp fabric reinforced cement nanocomposites during wet/dry cycles, particularly at 1 wt% CNC. In the final chapter, the results obtained in this work are summarised. The challenges and future research directions in this field are highlighted in the concluding remarks.

References

1. Neville AM. Properties of concrete. England: Pearson Education Limited; 2011.
2. Balaguru PN, Shah SP. Fiber-reinforced cement composites. USA: McGraw-Hill Inc; 1992.
3. Bentur A, Mindess S. Fibre reinforced cementitious composites. New York: Taylor & Francis; 2007.
4. Pacheco-Torgal F, Jalali S. Cementitious building materials reinforced with vegetable fibres: a review. Constr Build Mater. 2011;25:575–81.
5. Ardanuy M, Claramunt J, Toledo Filho RD. Cellulosic fiber reinforced cement-based composites: a review of recent research. Constr Build Mater. 2015;79:115–28.
6. Li Y, Mai YW, Ye L. Australia-Sisal fibre and its composites: a review of recent developments. Compos Sci Technol. 2000;60:2037–55.
7. Elsaid A, Dawood M, Seracino R, Bobko C. Mechanical properties of kenaf fiber reinforced concrete. Constr Build Mater. 2011;25:1991–2001.
8. Santos SF, Tonoli GHD, Mejia JEB, Fiorelli J, Savastano Jr H. Non-conventional cement-based composites reinforced with vegetable fibers: a review of strategies to improve durability. Materiales de Construcción. 2015;65. doi:10.3989/mc.2015.05514.
9. Silva FDA, Mobasher B, Filho RDT. Cracking mechanisms in durable sisal fiber reinforced cement composites. Cem Concr Compos. 2009;31:721–30.
10. Islam SM, Hussain RR, Morshed MAZ. Fiber-reinforced concrete incorporating locally available natural fibers in normal- and high-strength concrete and a performance analysis with steel fiber-reinforced composite concrete. J Compos Mater. 2011;46:111–22.
11. Hamzaoui R, Guessasma S, Mecheri B, Eshtiaghi AM, Bennabi A. Microstructure and mechanical performance of modified mortar using hemp fibres and carbon nanotubes. Mater Des. 2014;56:60–8.
12. Li Z, Wang L, Wang X. Compressive and flexural properties of hemp fiber reinforced concrete. Fibers Polym. 2004;5:187–97.
13. Misnon MI, Islam MM, Epaarachchi JA, Lau KT. Analyses of woven hemp fabric characteristics for composite reinforcement. Mater Des. 2015;66:82–92.
14. Li Z, Wang X, Wang L. Properties of hemp fibre reinforced concrete composites. Compos A. 2006;37:497–505.
15. Awwad E, Mabsout M, Hamad B, Farran MT, Khatib H. Studies on fiber-reinforced concrete using industrial hemp fibers. Constr Build Mater. 2012;35:710–7.
16. Peled A, Bentur A. Geometrical characteristics and efficiency of textile fabrics. Cem Concr Res. 2000;30:781–90.
17. Peled A, Sueki S, Mobasher B. Bonding in fabric–cement systems: effects of fabrication methods. Cem Concr Res. 2006;36:1661–71.
18. Mobasher B, Peled A, Pahilajani J. Distributed cracking and stiffness degradation in fabric-cement composites. Mater Struct. 2006;39:317–31.
19. Soranakom C, Mobasher B. Geometrical and mechanical aspects of fabric bonding and pullout in cement composites. Mater Struct. 2008;42:765–77.

20. Alamri H, Low IM, Alothman Z. Mechanical, thermal and microstructural characteristics of cellulose fibre reinforced epoxy/organoclay nanocomposites. Compos B. 2012;43:2762–71.
21. Alhuthali A, Low IM, Dong C. Characterisation of the water absorption, mechanical and thermal properties of recycled cellulose fibre reinforced vinyl-ester eco-nanocomposites. Compos B. 2012;43:2772–81.
22. Mohr BJ, Biernacki JJ, Kurtis KE. Microstructural and chemical effects of wet/dry cycling on pulp fiber–cement composites. Cem Concr Res. 2006;36:1240–51.
23. Almeida AEFS, Tonoli GHD, Santos SF, Savastano H. Improved durability of vegetable fiber reinforced cement composite subject to accelerated carbonation at early age. Cem Concr Compos. 2013;42:49–58.
24. Toledo Filho RD, Ghavami K, England GL, Scrivener K. Development of vegetable fibre–mortar composites of improved durability. Cem Concr Compos. 2003;25:185–96.
25. Soroushian P, Hassan M. Evaluation of cement-bonded strawboard against alternative cement-based siding products. Constr Build Mater. 2012;34:77–82.
26. Blankenhorn PR, Blankenhorn BD, Silsbee MR, Dicola M. Effects of fiber surface treatments on mechanical properties of wood fiber–cement composites. Cem Concr Res. 2001;31:1049–55.
27. Le Troëdec M, Peyratout CS, Smith A, Chotard T. Influence of various chemical treatments on the interactions between hemp fibres and a lime matrix. J Eur Ceram Soc. 2009;29:1861–8.
28. Rachini A, Le Troedec M, Peyratout C, Smith A. Comparison of the thermal degradation of natural, alkali-treated and silane-treated hemp fibers under air and an inert atmosphere. J Appl Polym Sci. 2009;112:226–34.
29. De Gutiérrez RM, Díaz LN, Delvasto S. Effect of pozzolans on the performance of fiber-reinforced mortars. Cem Concr Compos. 2005;27:593–8.
30. Mohr BJ, Biernacki JJ, Kurtis KE. Supplementary cementitious materials for mitigating degradation of kraft pulp fiber-cement composites. Cem Concr Res. 2007;37:1531–43.
31. Siddique R, Klaus J. Influence of metakaolin on the properties of mortar and concrete: a review. Appl Clay Sci. 2009;43:392–400.
32. Flatt RJ, Roussel N, Cheeseman CR. Concrete: an eco material that needs to be improved. J Eur Ceram Soc. 2012;32:2787–98.
33. Shaikh FUA. Review of mechanical properties of short fibre reinforced geopolymer composites. Constr Build Mater. 2013;43:37–49.
34. Juenger MCG, Siddique R. Recent advances in understanding the role of supplementary cementitious materials in concrete. Cem Concr Res. 2015;78:71–80.
35. Sanchez F, Sobolev K. Nanotechnology in concrete—a review. Constr Build Mater. 2010;24:2060–71.
36. Singh T. A review of nanomaterials in civil engineering works. Int J Struct Civil Eng Res. 2014;3:1–7.
37. Pacheco-Torgal F, Jalali S. Nanotechnology: advantages and drawbacks in the field of construction and building materials. Constr Build Mater. 2011;25:582–90.
38. Hanus MJ, Harris AT. Nanotechnology innovations for the construction industry. Prog Mater Sci. 2013;58:1056–102.
39. Singh LP, Karade SR, Bhattacharyya SK, Yousuf MM, Ahalawat S. Beneficial role of nanosilica in cement based materials—a review. Constr Build Mater. 2013;47:1069–77.
40. Supit SWM, Shaikh FUA. Effect of nano-$CaCO_3$ on compressive strength development of high volume fly ash mortars and concretes. J Adv Concr Technol. 2014;12:178–86.
41. Chang TP, Shih JY, Yang KM, Hsiao TC. Material properties of portland cement paste with nano-montmorillonite. J Mater Sci. 2007;42:7478–87.
42. Morsy MS, Aglan HA, Abd El Razek MM. Nanostructured zonolite–cementitious surface compounds for thermal insulation. Constr Build Mater 2009;23:515–21.
43. Wei J, Meyer C. Degradation mechanisms of natural fiber in the matrix of cement composites. Cem Concr Res. 2015;73:1–16.

44. Farzadnia N, Abang Ali AA, Demirboga R, Anwar MP. Effect of halloysite nanoclay on mechanical properties, thermal behavior and microstructure of cement mortars. Cem Concr Res. 2013;48:97–104.
45. He C, Makovicky E, Osbaeck B. Thermal treatment and pozzolanic activity of Na and Ca-montmorillonite. Appl Clay Sci. 1996;10:351–68.
46. Shebl SS, Allie L, Morsy MS, Aglan HA. Mechanical behavior of activated nano silicate filled cement binders. J Mater Sci. 2009;44:1600–6.
47. Morsy MS, Alsayed SH, Aqel M. Effect of nano-clay on mechanical properties. Int J Civil Environ Eng. 2010;10:23–7.

Chapter 2
Materials and Methodology

2.1 Materials

Hemp fabric (HF) and nanoclay platelets (Cloisite 30B) were used as reinforcements for the cement-matrix composites in this research. The hemp fabric was supplied by Hemp Wholesale Australia, Kalamunda, Western Australia as shown in Fig. 2.1. The chemical composition and also the physical properties and structure of hemp fabric are shown in Tables 2.1 and 2.2 respectively. The nanoclay platelets (Cloisite 30B) used in this investigation are based on natural montmorillonite clay (hydrated sodium calcium aluminium magnesium silicate hydroxide ($Na,Ca)_{0.33}(Al, Mg)_2(Si_4O_{10})(OH)_2 \cdot nH_2O$). Cloisite 30B is a natural montmorillonite modified with a quaternary ammonium salt, which was supplied by Southern Clay Products, USA. The specification and physical properties of Cloisite 30B are outlined in Table 2.3. Ordinary Portland cement (OPC) was used in all mixes. The chemical composition and physical properties of OPC are listed in Table 2.4. Calcined nanoclay (CNC) was prepared by heating the nanoclay at 800, 850 and 900 °C for 2 h in an electric furnace with a heating rate of 10 °C/min. The calcined nanoclay was then characterized by XRD and TEM in order to determine the amorphous phase of calcined nanoclay at calcination temperature. In order to treat the surface of the fibres, the hemp fabrics were immersed in 1.7 M NaOH solution (pH = 14) for 48 h at 25 °C and then neutralized with 1% vol. acetic acid. They were then washed several times with deionized water until the pH reached about at 7. Finally the fabrics were dried in an oven at 40 °C for 24 h.

© The Author(s) 2017
I.-M. Low et al., *High Performance Natural Fiber-Nanoclay Reinforced Cement Nanocomposites*, Biobased Polymers, DOI 10.1007/978-3-319-56588-0_2

Fig. 2.1 Optical micrograph
of hemp fabric [1]

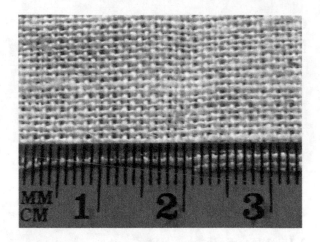

Table 2.1 Chemical analysis of hemp [2]

	Cellulosic residue (wt%)	Pectin (wt%)	Hemicellulose (wt%)	Lignin (wt%)	(Wax, fat, protein) (wt%)
Hemp fibre	56.1	20.1	10.9	6	7.9

Table 2.2 Properties and structure of hemp fabric [2, 3]

Fabric thickness (mm)	0.43
Fabric geometry	Woven (plain weave)
Yarn nature	Bundle
Filament size (mm)	0.04253
Number of filaments in a bundle	24
Bundle diameter (mm)	0.21
Opening size (mm)	0.3
Fabric density (gm/cm^3)	0.6
Modulus of elasticity (GPa)	38–58
Tensile strength (MPa)	591–857

Table 2.3 Physical properties of the nanoclay platelets (Cloisite 30B) [4]

Physical properties of the (Cloisite 30B)	
Colour	Off white
Density (g/cm^3)	1.98
d-spacing (001) (nm)	1.85
Aspect ratio	200–1000
Surface area (m^2/g)	750
Mean particle size (μm)	6

Table 2.4 Physical properties and chemical composition of OPC [5]

Properties/Compositions	OPC (ASTM Type I)
Physical properties	
Specific gravity	3.17
Specific surface, Blaine (cm^2/gm)	3170
Chemical analysis	
SiO_2	21.10
Al_2O_3	5.24
Fe_2O_3	3.10
CaO	64.39
MgO	1.10
SO_3	2.52
Na_2O	0.23
K_2O	0.57
LOI	1.22

2.2 Sample Preparation

2.2.1 Cement Nanocomposite

Ordinary Portland cement (OPC) is partially substituted by nanoclay (NC) or calcined nanoclay (CNC) of 1, 2 and 3% by weight of OPC. The OPC and NC or CNC were first dry mixed for 5 min in a Hobart mixer at a low speed and then mixed for another 10 min at high speed until homogeneity was achieved. The binder is either nanoclay-cement dry powder or calcined nanoclay-cement dry powder. The cement nanocomposite paste was prepared through adding water with a water/binder ratio of 0.485. The cement nanocomposite containing 1, 2 and 3 wt% NC is termed as NCC1, NCC2 and NCC3, respectively. And also the cement nanocomposite containing 1, 2 and 3 wt% calcined NC is termed as CNCC1, CNCC2 and CNCC3, respectively. The cement paste (C) was considered as a control. The mix proportions are shown in Table 2.5.

Table 2.5 Mix proportions of specimens in Chap. 3

Sample name	Hemp fabric (HF) (wt%)	Mix proportions (wt%)		
		Cement	Nanoclay	Water/binder
NCC-0	0	100	0	0.485
NCC-1	0	99	1	0.485
NCC-2	0	98	2	0.485
NCC-3	0	97	3	0.485
NCC-0/HF	2.5	100	0	0.485
NCC-1/HF	2.5	99	1	0.485
NCC-2/HF	2.5	98	2	0.485
NCC-3/HF	2.5	97	3	0.485

2.2.2 Untreated and Treated Hemp Fabric-Reinforced Cement Nanocomposites

The cement paste (C) was prepared through adding water (W) with W/C ratio of 0.485. The fabrication of untreated hemp fabric-reinforced cement composite (UHFRC) samples was done in two stages. In the first stage, the hemp fabric (295 mm in length and 65 mm in width) was first soaked into the cementitous matrix in order to achieve better penetration of matrix into the openings of the fabric. Then layers of pre-soaked hemp fabric were laid on a polished timber plate. The compacted layers of fabric were then left under a 30 kg weight (or 4.9 kPa compressive pressure) for 1 h to reduce the air bubbles and voids which might otherwise be trapped inside the samples. This step is essential to ensure better penetration of the cement matrix into the filaments of the hemp fabric and thus improves the interfacial bonding between the fibre and the matrix. In the second stage, a thin layer of cement matrix was first poured into the prismatic mould followed by the compacted pre-soaked hemp fabrics. Finally a thin layer of matrix was poured into the mould to form the upper layer and the samples were left to cure for 24 h at room temperature. Samples of un-treated hemp fabric-reinforced cement composites were fabricated with various contents of hemp fabric: 4.5 wt% (4 layers of fabric), 5.7 wt% (5 layers of fabric), 6.9 wt% (6 layers of fabric) and 8.1 wt% (7 layers of fabric). For the fabrication of treated hemp fabric-reinforced cement composite (6THFRC) samples, only 6 layers of treated hemp fabric were used because they have been shown to exhibit the best mechanical performance. The fabrication procedure of 6THFRC was similar to that of UHFRC described above. A schematic of of 6 treated hemp fabric layers position through the depth of sample is shown in Fig. 2.2. The mix proportions are shown in Table 2.6.

2.2.3 Thermal Treatment of Nanoclay

Calcined nanoclay (CNC) was prepared by heating the nanoclay at 800, 850 and 900 °C for 2 h in an electric furnace with a heating rate of 10 °C/min. The calcined

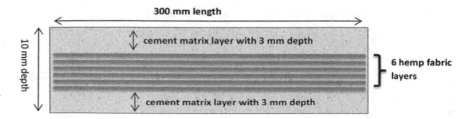

Fig. 2.2 Schematic representation of 6 treated hemp fabric layers position through the depth of sample [1]

Table 2.6 Mixing proportions of specimens in Chap. 5

Sample	Hemp fabric (HF)		Mix proportions (wt%)		
	Content (wt%)	Fabric layers	Cement	CNC	Water/binder
C	0	0	100	0	0.485
4UHFRC	4.5	4	100	0	0.485
5UHFRC	5.7	5	100	0	0.485
6UHFRC	6.9	6	100	0	0.485
7UHFRC	8.1	7	100	0	0.485
6THFRC	6.9	6	100	0	0.485
CNCC1	0	0	99	1	0.485
CNCC2	0	0	98	2	0.485
CNCC3	0	0	97	3	0.485
6THFR-CNCC1	6.9	6	99	1	0.485
6THFR-CNCC2	6.9	6	98	2	0.485
6THFR-CNCC3	6.9	6	97	3	0.485

nanoclay was then characterized by XRD, EDS and TEM in order to determine the amorphous phase of calcined nanoclay.

2.2.4 Curing and Specimens

For each series, five prismatic plate specimens of dimensions 300 mm × 70 mm × 10 mm were cast. All specimens were demolded after 24 h of casting and kept under water for about 56 days. For durability test, the period of the wetting and drying cycles was determined as 30 days under water followed by 30 days of drying in air for one cycle and it was performed for 3 cycles, after that the samples tested at 236 days counting from the casting day. Five rectangular specimens of each series with dimensions 70 mm × 20 mm × 10 mm were cut from the fully cured prismatic plate for each mechanical and physical test at 56 and 236 days.

2.3 Material Characterisation

2.3.1 X-Ray Diffraction

The samples were measured on a D8 Advance Diffractometer (Bruker-AXS) using copper radiation and a LynxEye position sensitive detector. The diffractometer were scanned from 7° to 70° (2θ) in steps of 0.015° using a scanning rate of 0.5°/min. XRD patterns were obtained by using Cu Ka lines (λ = 1.5406 Å). A knife edge collimator was fitted to reduce air scatter.

The Quantitative X-ray Diffraction Analysis (QXDA) with Rietveld refinement was done with Bruker *DIFFRAC^{plus}* EVA software associated with the International Centre for Diffraction Data PDF-4 2013 database. Corundum [Al_2O_3] was chosen to serve as an internal standard. It was selected because it does not overlap with important cement peaks up to 2θ of 60° as well as it does not react with water and has no influence on the hydration reaction [6–11]. By using an internal standard the concentration of the crystalline phase can be determined on an absolute basis enabling the amorphous fraction to also be determined. The samples for QXDA were prepared by mixing a dry weight of 3.0 g of cement paste or nano-composite with 0.33 g of Corundum [Al_2O_3] as the internal standard. This powder was then added to a McCrone micronising canister with 7 ml of laboratory grade ethanol and sintered alumina milling media and milled for 5.0 min. The suspension was then poured into a polypropylene dish and dried at 105 °C for 24 h. The dried powder was then brushed into a polypropylene vial, and sealed until analysis [12].

2.3.2 High Resolution Transmission Electron Microscopy

High Resolution Transmission electron microscopy imaging was done using 3000F (JEOL company) operating at 300 kV equipped with a 4 × 4 k CCD camera (Gatan). HREM is an imaging technique that creates images with atomic resolution. 3000F has excellent HREM performance including 0.195 nm point resolution and 0.104 nm lattice resolution. HRTEM was carried out at University of Western Australia. Nanoclay (Cloisite 30B) powder was dispersed in ethanol inside small glass container by using ultrasonic device for 15 min. After that few drops of suspension were mounted onto copper grid and then kept to dry.

2.3.3 Synchrotron Radiation Diffraction

Synchrotron radiation diffraction (SRD) measurement was carried out on the powder diffraction beamline at the Australian Synchrotron. The diffraction patterns of each sample were collected using a wavelength of 0.825 Å in the two-theta range of 8–52°.

2.3.4 Scanning Electron Microscopy

Scanning electron microscopy imaging was obtained using a NEON 40ESB, ZEISS. The SEM investigation was carried out in detail on microstructures and the fractured surfaces of samples. Specimens were coated with a thin layer of platinum before observation by SEM to avoid charging.

2.3.5 Thermogravimetric Analysis

The thermal stability of samples was studied by thermogravimetry analysis (TGA). A Mettler Toledo TGA 1 star system analyser was used for all these measurements. Samples with 25 mg were placed in an alumina crucible and tests were carried out in Argon atmosphere with a heating rate of 10 °C/min from 25 to 1000 °C.

2.4 Physical Properties

Measurements of bulk density and porosity were conducted to determine the quality of of nanocomposites and HF-reinforced nanocomposites accordance with the ASTM Standard (C-20) [13]. The thickness, width, length and weight are measured in order to determine the bulk density. The calculation for density was carried out by using the following equation:

$$\rho = \frac{m_d}{V} \tag{2.1}$$

where, ρ = density in (g/cm^3), m_d = mass of the dried sample (g) and V = volume of the test specimen (cm^3).

The value of apparent porosity P_S was determined using the Archimedes principle in accordance with the ASTM Standard (C-20) and clean water was used as the immersion water. The apparent porosity P_S was calculated using the following equation [13]:

$$P_S = \frac{m_s - m_d}{m_s - m_i} \times 100 \tag{2.2}$$

where m_i = mass of the sample saturated with and suspended in water, m_s = mass of the sample saturated in air.

For the water absorption test, the produced specimens were dried at a temperature of 80 °C until their mass became constant and then the mass was weighed (W_0). The specimens were then immersed in clean water at a temperature of 20 °C for 48 h. After the desired immersion period, the specimens were taken out and wiped quickly with wet cloth, and then the mass was weighed (W_1) immediately. The rate of water absorption (W_A) was calculated by using the formula:

$$W_A = \frac{W_1 - W_0}{W_0} \times 100 \tag{2.3}$$

2.5   Mechanical Properties

2.5.1   Compressive Strength

Compressive strength of specimens was tested according to ASTM: C109 using a loading rate of 0.33 MPa/s. The cube samples of size 50 × 50 × 50 mm are cast. Five cubic specimens of each composition were used to measure the compressive strength.

2.5.2   Flexural Strength and Fracture Toughness

Three-point bend tests were conducted using a LLOYD Material Testing Machine to evaluate the flexural strength and fracture toughness of the specimens. The support span used was 40 mm with a displacement rate of 0.5 mm/min. The flexural strength σ_F was evaluated using the following equation:

$$\sigma_F = \frac{3P_mS}{2BW^2} \qquad (2.4)$$

where P_m is the maximum load, S is the span of the sample, W is the specimen depth and B is the specimen width.

In order to determine the fracture toughness, a sharp razor blade was used to initiate a sharp crack in the samples. The ratio of crack length to depth ($\frac{a}{W}$) was about 1/3. The fracture toughness was calculated using the following equation [14, 15]:

$$K_{IC} = \frac{P_mS}{BW^{3/2}}f\left(\frac{a}{W}\right) \qquad (2.5a)$$

where a is the crack length (mm) and $f(\frac{a}{W})$ is the polynomial geometrical correction factor given by:

$$f(\frac{a}{W}) = \frac{3(a/W)^{1/2}[1.99 - (a/W)(1 - a/W) \times (2.15 - 3.93a/W + 2.7a^2/W^2)]}{2(1 + 2a/W)(1 - a/W)^{3/2}}$$

$$(2.5b)$$

Five specimens, measuring 70 × 20 × 10 mm, of each composition were used to measure the flexural strength and the fracture toughness.

2.5.3 Impact Strength

The impact strength of the specimen was determined using a Zwick Charpy impact tester with 15 J pendulum hammer and 40 mm support span. Un-notched samples were used to compute the impact strength using the following formula:

$$\sigma_I = \frac{E}{A} \tag{2.6}$$

where E is the impact energy to break a sample with a ligament of area A. Five specimens, measuring $70 \times 20 \times 10$ mm, of each composition were used to measure the impact strength.

References

1. Hakamy A. Microstructural design of high-performance natural fibre nanoclay cement nanocomposites. Ph.D. Thesis. Perth Australia: Curtin University; 2016.
2. Sedan D, Pagnoux C, Smith A, Chotard T. Mechanical properties of hemp fibre reinforced cement: influence of the fibre/matrix interaction. J Eur Ceram Soc. 2008;28:183–92.
3. Peled A, Sueki S, Mobasher B. Bonding in fabric–cement systems: effects of fabrication methods. Cem Concr Res. 2006;36:1661–71.
4. Alhuthali A, Low IM, Dong C. Characterization of the water absorption, mechanical and thermal properties of recycled cellulose fibre reinforced vinyl-ester eco-nanocomposites. Compos Part B. 2012;43:2772–81.
5. Ahmed SFU, Maalej M, Paramasivam P. Flexural responses of hybrid steel-polyethylene fiber reinforced cement composites containing high volume fly ash. Constr Build Mater. 2007;21:1088–97.
6. ASTM C-1365 – 06. Standard test method for determination of the proportion of phases in Portland cement and Portland-cement clinker using X-ray powder diffraction analysis. ASTM International; 2011.
7. Taylor HFW. Cement chemistry. London: Academic press limited; 1990.
8. Aldridge A. Accuracy and precision of phase analysis in Portland cement by Bogue, microscopic and x-ray diffraction methods. Cem Concr Res. 1982;12(3):381–98.
9. Suherman P, Riessen A, O'Connor B, Li D, Bolton D, Fairhurst H. Determination of amorphous phase levels in Portland cement clinker. Powder Diffr. 2002;17(3):178.
10. Scrivenera K, Fullmanna T, Galluccia E, Walentab G, Bermejob E. Quantitative study of Portland cement hydration by X-ray diffraction/Rietveld analysis and independent methods. Cem Concr Res. 2004;34(9):1541–7.
11. Torre G, Bruque S, Aranda M. Rietveld quantitative amorphous content analysis. J Appl Crystallogr. 2001;34(2):196–202.
12. Williams R, Riessen A. Determination of the reactive component of fly ashes for geopolymer production using XRF and XRD. Fuel. 2010;89(12):3683–92.
13. ASTM C-20. Standard test methods for apparent porosity, water absorption, apparent specific gravity, and bulk density of burned refractory brick and shapes by boiling water. ASTM International; 2010.
14. ASTM E-399. Standard fracture toughness specimens. ASTM International; 2013.
15. Mihashi H, de Barros Leite JP, Yamakoshi S, Kawamata A. Controlling fracture toughness of matrix with mica flake inclusions to design pseudo-ductile fibre reinforced cementitious composites. Eng Fract Mech. 2007;74:210–22.

Chapter 3
Hemp Fabric Reinforced Organoclay–Cement Nanocomposites: Microstructures, Physical, Mechanical and Thermal Properties

3.1 Introduction

This chapter presents the micro-structural characterisation, physical, mechanical and thermal properties of nanoclay-cement nanocomposites fabricated and tested at 56 days. The influence of nanoclay on properties of hemp fabric-reinforced cement composites is also presented in this chapter. Two layers of hemp fabric (HF) were used and the total amount of hemp fabric in each specimen was about 2.5 wt%. In this study eight series of composites are considered. The first series is the control and made of 100% Ordinary Portland cement without any Hemp fabric and is termed as NCC-0. The second, third and fourth series are similar to the first series in every aspect except where the OPC was partially substituted by nanoclay with 1, 2 and 3 wt% of OPC. They are termed as NCC-1, NCC-2 and NCC-3, respectively. In fifth series, two layers of HF which is equivalent to 2.5 wt% is added to reinforce the cement matrix to fabricate HF reinforced cement composite, which is termed as NCC-0/HF. The sixth, seventh and eighth series are similar to fifth series in every aspect except where the OPC was partially substituted by nanoclay with 1, 2 and 3 wt% of OPC. They are termed as NCC-1/HF, NCC-2/HF and NCC-3/HF, respectively. Detail mix proportions of above series can be found in Table 2.5 of Chap. 2. For each series, three prismatic plate specimens of 300 mm × 70 mm × 10 mm were cast then after 56 days several rectangular specimens of each series with dimensions 70 mm × 20 mm × 10 mm were cut from the fully cured prismatic plate for mechanical and physical tests. The results indicated that hemp fabric-reinforced nanocomposites containing 1 wt% nanoclay had denser microstructure than others, and thus this improvement led to enhance the hemp fabric-nanomatrix adhesion. In addition, the incorporation of 1 wt% nanoclay into the hemp fabric-reinforced nanocomposites improved the thermal stability, decreased the porosity and water absorption as well as increased the density, flexural strength, fracture toughness, impact strength and Rockwell hardness as well as improved thermal stability when compared to the hemp fabric-reinforced cement composite.

© The Author(s) 2017
I.-M. Low et al., *High Performance Natural Fiber-Nanoclay Reinforced Cement Nanocomposites*, Biobased Polymers, DOI 10.1007/978-3-319-56588-0_3

3.2   Results and Discussion

3.2.1   High Resolution Transmission Electron Microscopy

HRTEM images for nanoclay (Cloisite 30B) are shown in Fig. 3.1a, b. The lower magnification image in Fig. 3.1a gives a general view of the nanoclay platelets. The high magnification image in Fig. 3.1b shows the layer structure of nanoclay platelets. It can be seen clearly that the distances between the nanoclay platelets were about 1.85 nm and thus this is evidence that the d-spacing of (0 0 1) planes in nanoclay layers were 1.85 nm as shown in Table 2.3 in Chap. 2.

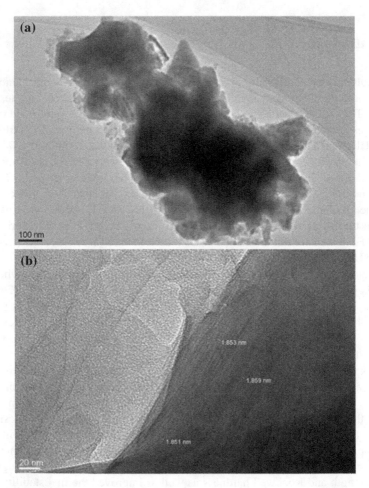

Fig. 3.1   TEM images of nanoclay (Cloisite 30B) at: **a** low magnification, **b** high magnification [1]

3.2.2   XRD Analysis

Figure 3.2a–d show the XRD patterns of nanoclay, control cement paste and nanocomposites containing 1 and 3 wt% nanoclay respectively. International Centre for Diffraction Data (PDF-4 2013) database was used for phase identification. Figure 3.2a shows XRD patterns of nanoclay, a broader diffraction peaks such as at 2θ of $20.0°$, indicate that nanoclay is mostly amorphous substance having a poor crystalline structure. However, nanoclay perhaps has two trivial crystalline phases such as Montmorillonite-18A $[Na_{0.3}(Al,Mg)_2Si_4O_{10}OH_2 \cdot 6H_2O]$ (PDF000120219) and Illite-montmorillonite (NR) $[KAl_4(Si,Al)_8O_{10}(OH)_4 \cdot 4H_2O]$ (PDF 000350652).

Fig. 3.2 XRD patterns of: **a** nanoclay, **b** control cement paste, **c** nanocomposites containing 1 wt% nanoclay, **d** nanocomposites containing 3 wt% nanoclay [1]

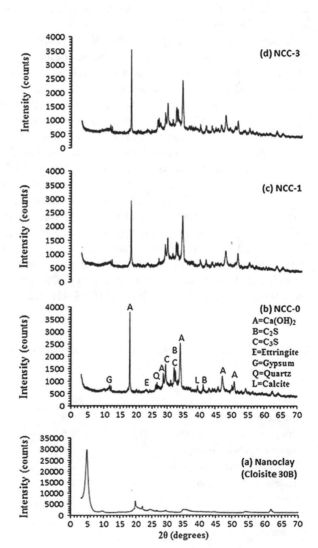

In Fig. 3.2b–d, there are three important phases in this study: portlandite [Ca (OH)$_2$] (PDF 00-044-1481), dicalcium silicate [C$_2$S] (PDF 00-033-0302) and tricalcium silicate [C$_3$S] (00-049-0442). Moreover, there are four less important phases: Ettringite [Ca$_6$Al$_2$(SO$_4$)$_3$(OH)$_{12}$·26H$_2$O] (PDF 000411451), Gypsum [Ca (SO$_4$)(H$_2$O)$_2$] (PDF 040154421), Quartz [SiO$_2$] (PDF 000461045) and Calcite [CaCO$_3$] (PDF 000050586). The composition of Ca(OH)$_2$ has a well-defined crystallized structure, it has five major peaks in the XRD pattern that corresponds to 2θ of 18.01°, 28.67°, 34.10°, 47.12° and 50.81°, and crystal planes of Miller indices (hkl) of (001), (100), (101), and (102) and (110), respectively. Although there are some overlaps of peaks and they have small intensities, dicalcium silicate (C$_2$S) has four major peaks that correspond to 2θ of 32.05°, 32.14°, 32.59° and 41.21° as well as tricalcium silicate (C$_3$S) has four major peaks that correspond to 2θ of 29.29°, 32.12°, 32.46° and 51.75°.

Table 3.1 shows the Quantitative X-ray Diffraction Analysis (QXDA) with Rietveld refinement of cement paste and nanocomposites containing 1 and 3 wt% nanoclay, respectively. Corundum [Al$_2$O$_3$] was chosen to serve as an internal standard. It was selected because it does not overlap with important cement peaks up to 2θ of 60° as well it does not react with water and has no influence on the hydration reaction. However, generally, the addition of 1 and 3 wt% nanoclay into the cement matrix has resulted in apparent change to the crystalline components of the samples. As can be seen from Table 3.1 and Fig. 3.2c, the addition of 1 wt% nanoclay reduced the amount of Ca(OH)$_2$ from 19.5 to 15.7%, about 19.5% reduction compared to cement paste. The intensity Ca(OH)$_2$ at 2θ angle of 18.01° also decrease from 3817 to 2940, about 23% reduction compared to cement paste and also other major peaks of Ca(OH)$_2$ crystals were significantly reduced (Fig. 3.2b and c). Furthermore, the amorphous content was increased slightly from 67.4 to 70%, about 3.7% increase. This result indicates that an obvious consumption of Ca (OH)$_2$ crystals mainly due to the effect of pozzolanic reaction in the presence of nanoclay and good dispersion of nanoclay in the matrix leads to produce more amorphous calcium silicate hydrate gel (C–S–H). This explanation is also confirmed

Table 3.1 QXDA results showing the phase abundances in cement paste and nanocomposites containing 1 and 3 wt% nanoclay [1]

Weight %			
Phase	NCC-0	NCC-1	NCC-3
Portlandite [Ca(OH)$_2$]	19.5	15.7	18.6
Ettringite [Ca$_6$Al$_2$(SO$_4$)$_3$(OH)$_{12}$·26H$_2$0]	1.8	1.1	1.2
Gypsum [Ca(SO$_4$)(H$_2$O)$_2$]	0.4	0.4	0.6
Tricalcium silicate [C$_3$S]	1.5	2.1	1.6
Dicalcium silicate [C$_2$S]	7.1	8.5	7.4
Quartz [SiO$_2$]	0.4	0.6	0.6
Calcite [CaCO$_3$]	1.1	0.8	1.2
Amorphous content	67.4	70.0	68.0

by the inspection of amount of unreacted C_3S (2.1%) and C_2S (8.5), in which the amount of unreacted C_3S and C_2S are slightly higher than the cement paste. Wei et al. [2] reported that pozzolanic reaction decelerates the hydration reaction of C_3S and C_2S during the curing time of 28–112 days. In this study, these unreacted phases could react with water later to produce more C–S–H gel after 56 days.

On the other hand, as can be seen from Table 3.1 and Fig. 3.2d for nanocomposites containing 3 wt% nanoclay, there is insignificant effect of nanoclay in the hydration reaction. For example, the amount of $Ca(OH)_2$ was reduced from 19.5 to 18.6%, about 4.6% reduction compared to cement paste. The intensity of $Ca(OH)_2$ at 2θ angle of 18.01° was also slightly reduced from 3817 to 3549, about 7% reduction compared to cement paste (Fig. 3.2b and d). This may be attributed to agglomerations of nanoclay at high contents which lead to poor dispersion of nanoclay and hence poor pozzolanic reaction. Moreover, the amount of C_3S (1.6%) is similar to the cement paste, and the amount of C_2S (7.4%) is slightly higher than the cement paste by about 4.2%; this also confirms that hydration reaction has occurred more than pozzolanic reaction. Table 3.1 shows that Ettringite is slightly less in nanocomposites than cement paste. For example, it decreased from 1.8 to 1.1% in nanocomposite containing 1% nanoclay. Overall, the results indicate than 1 wt% nanoclay can consume more $Ca(OH)_2$ crystals and can improve the structure more effectively than 3 wt% nanoclay. The mechanical and physical properties also confirm this phenomenon, which is discussed in the following sections.

3.2.3   Porosity and Density

The porosity and density values of cement paste, nano-composites, HF reinforced cement paste and HF reinforced nanocomposites are shown in Fig. 3.3. Generally the composites containing HF exhibited higher porosity that these without HF. This could be attributed to the formation of voids at the interfacial areas between HF and matrices.

However, Fig. 3.3 shows that the addition of nanoclay decreases the porosity of these composites when compared to control cement paste and HF composites. For nanocomposites with 1 wt% of nanoclay, the porosity decreases by 20.6%. Moreover, in HF reinforced nanocomposites with 1 wt% of nanoclay, it decreases by 16%. This indicates that nanoclay has filling effect in the porosity of cement paste composites with and without HF. This result is in agreement with the work done by Jo et al. [3] where the porosity of cement mortar is decreased by the addition of nano-SiO_2 particles. The addition of 1 wt% of nanoclay increased the density of control cement paste and HF nano-composites by 4 and 3.4% respectively. That improvement demonstrated that cement composites with 1 wt% nanoclay yields more consolidated microstructure. However, the addition of more nanoclay leads to increase in porosity and decrease in density [4]. These results are also supported by the SEM examinations for the microstructure of cement paste, nanocomposites containing 1 and 3 wt% nanoclay. Figure 3.4a shows the deposits

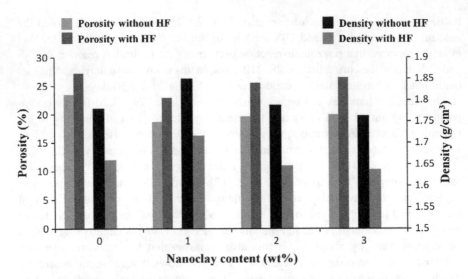

Fig. 3.3 Porosity as a function of nanoclay content for control cement and its nanocomposites with and without HF [1]

of Ca(OH)$_2$ crystals are distributed in the cement paste and also in Fig. 3.4b, the C-S-H gel connected with many ettringite (needle like hydration products), as well as the pores were extremely existed, in which that revealed weak structure. Figure 3.4c shows the SEM micrograph for nanocomposites containing 1 wt% nanoclay, it is different from that of cement paste, the structure is denser and compact with few pores. On the other hand, in Fig. 3.4d, the nanocomposites containing 3 wt% nanoclay shows more pores and microcracks which weaken the structure.

3.2.4   Mechanical Properties

In general, the addition of nanoclay improved the mechanical properties of the cement matrix. In addition, the pre-soaking of hemp fabric in cement paste during sample preparation leads to good penetration of the cement matrix in between the reinforced filaments of the bundle as shown in Fig. 3.5, which also improved the mechanical properties of samples.

3.2.5   Flexural Strength

Figures 3.6 and 3.7 show flexural strength and load-midspan deflection curves for control cement paste, nano-composites, HF reinforced cement paste and HF

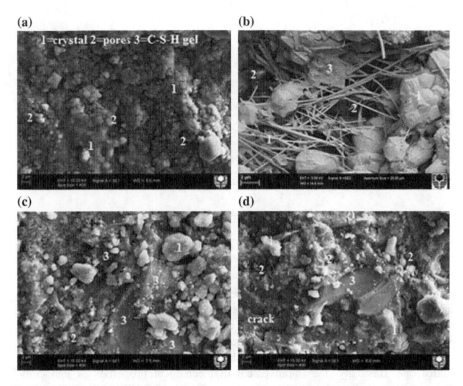

Fig. 3.4 SEM micrographs of: **a** and **b** cement paste, **c** nanocomposites containing 1 wt% nanoclay and **d** nanocomposites containing 3 wt% nanoclay [1]

Fig. 3.5 Penetration of the cement matrix in between the reinforced fibres of hemp fabric [1]

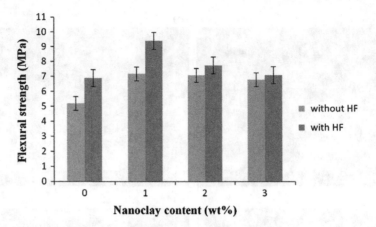

Fig. 3.6 Flexural strength as a function of nanoclay content for control cement paste and its nanocomposites with and without HF [1]

reinforced nano-composites. In general, the incorporation of nanoclay platelets into cement matrix led to a modest enhancement in flexural strength for all nanocomposite samples as shown in Fig. 3.6. The addition of 1% nanoclay resulted in the highest flexural strength of all nanocomposites. The flexural strength of nanocomposites containing 1 wt% nanoclay is increased by 38.1% compared to control one. Basically, the improvement in mechanical properties of nanocomposites can be attributed to two mechanisms. First is the filling effect where the nanoclay filled the voids or pores in cement paste in which the nano-particles were uniformly dispersed in the matrix thus making the microstructure of nanocomposites denser than the control cement paste [5, 6]. However, the addition of more nanoclay than 1 wt% caused a marked decrease in flexural strength. This can be due to the poor dispersion and agglomerations of the nanoclay in the cement matrix at higher clay contents, which create weak zones, in form of micro-voids as stress concentrators [5, 7].

The effect of nanoclay on the flexural strength of HF reinforced cement composite can also be seen in Fig. 3.6. The flexural strength of HF reinforced nanocomposites containing 1 wt% nanoclay is increased from 6.88 to 9.37 MPa, about 38.1% increase compared to HF reinforced cement composite. This improvement is explained as follows: the degradation of natural fibres in Portland cement matrix is due to the high alkaline environment (calcium hydroxide solution) which dissolves the lignin and hemicellulose phases, thus weakening the fibre structure [8]. In order to improve the durability of natural fibre in cement paste, the matrix could be modified in which calcium-hydroxide (CH) must be mostly consumed or reduced [9]. In this study, the nanoclay effectively prevented hemp fabric degradation through pozzolanic reaction as described above. Thus, the degradation of hemp fibres in nanocomposite is mostly prevented, especially in the case of using 1 wt% nanoclay, which is evident from the higher flexural strength value. However,

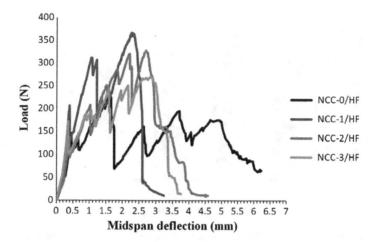

Fig. 3.7 Load-midspan deflection curves for curves for HF reinforced cement composite and HF reinforced nanocomposites from flexural test [1]

the addition of more nanoclay (i.e. 3 wt%) led to a slight reduction in flexural strength due to increase in porosity.

The load-midspan deflection curves for HF reinforced cement composite and HF reinforced nanocomposites are shown in Fig. 3.7. The HF reinforced nanocomposites containing 1 wt% nanoclay shows highest flexural load but smaller deflection capacity at peak load. This is due to high fibre-matrix interface bond, which increases the maximum load capacity. On the other hand, the HF reinforced nanocomposites containing 2 and 3 wt% nanoclay and HF reinforced cement composite show low flexural load but large deflection capacity at peak load, thereby higher ductility. This could be attributed to the increase in porosity which decreases the bond strength of fibre-matrix adhesion.

3.2.6 Flexural Modulus

The calculated flexural modulus of all composites is shown in Fig. 3.8 which shows that the addition of 1 wt% nanoclay in cement matrix increased the flexural modulus by 18.9% over control. The addition of HF fabrics adversely affected the flexural modulus for control cement composite and nanocomposites. For example, flexural modulus of control cement paste was decrease from 802.04 to 598.42 MPa, about 25.4% decrease, after the addition of HF fabrics. Similarly, Elsaid et al. [10] reported that as the length of the natural fibres used in a concrete mixture increase, the flexural modulus decrease. However, in HF reinforced nanocomposites the flexural modulus was slightly enhanced. For instance, flexural modulus of HF reinforced nanocomposites containing 1 wt% nanoclay was increase from 802.04 to 876.93 MPa, about 8.5% increase compared to control cement paste.

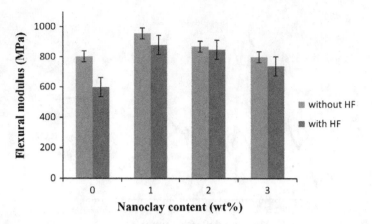

Fig. 3.8 Flexural modulus as a function of nanoclay content for control cement paste and its nanocomposites with and without HF [1]

3.2.7 *Fracture Toughness*

The effect of nanoclay on fracture toughness of control and nanocomposites without and with HF is shown in Fig. 3.9. In general, all composites containing HF showed significantly improvement in fracture toughness as well as multiple cracking behaviour as shown in Fig. 3.10 [11]. In case of nanocomposites, the fracture toughness of control and nanocomposites with 1, 2 and 3 wt% nanoclay was 0.356, 0.492, 0.421 and 0.396 MPa m$^{1/2}$, respectively. It can be seen clearly that, the nanocomposites with 1 wt% of nanoclay achieve better fracture properties with improvement reaching up to 38.2%. In similar study, Alamri and Low [12] reported

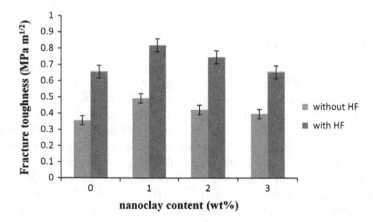

Fig. 3.9 Fracture toughness as a function of nanoclay content for control cement paste and its nanocomposites with and without HF [1]

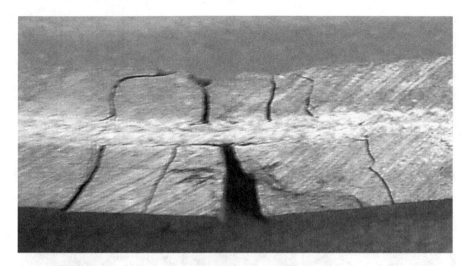

Fig. 3.10 SEM micrographs of failure mechanisms [1]

that addition of 1 wt% nano-SiC into epoxy matrix significantly increased the fracture toughness by a maximum 89.4% compared to neat epoxy. However, Fig. 3.9 also shows that facture toughness of nanocomposites decreased slightly when more nanoclay was added. Because the poor dispersion of high content of nanoclay leads to agglomeration which increased the porosity as observed in this study. Moreover, agglomeration acts as a stress concentration that can initiate tiny cracks, which leads to crack propagation [5, 12].

As expected, the composites reinforced with HF showed a significant increase in fracture toughness. This extraordinary enhancement is due to fracture resistance by hemp fabrics which resulted in increased energy dissipation from crack-deflection at the fibre-matrix interface, fibre-debonding, fibre-bridging, fibre pull-out and fibre-fracture [10, 11, 13, 14].

The addition of nanoclay, in HF reinforced nanocomposites also increased the fracture toughness. The fracture toughness for HF reinforced nanocomposites containing 1, 2 and 3 wt% nanoclay is 0.818, 0.744, and 0.652 MPa $m^{1/2}$, respectively. This is attributed to the fact that the nanoclay prevents hemp fibre degradation through pozzolanic reaction and reduction of CH. Thus, in this study the hemp fibres are mostly prevented, especially in the case of 1 wt% nanoclay in which the increase in fracture toughness was 129.8% comparing to control cement matrix. However, the inclusion of high content of nanoclay increases the porosity which adversely affected the interfacial bond between the matrix and the fibres thus decreased the fracture toughness [15].

Figure 3.11 shows the SEM micrographs of the fracture surface and HF/matrix interface of HF reinforced cement composite and HF reinforced nanocomposite containing 1 and 3 wt% nanoclay after fracture toughness test. A variety of toughness mechanisms such as shear deformation, crack bridging, fibre pull-out and

Fig. 3.11 SEM images of the fracture surface at low and high magnification: **a** and **b** HF reinforced cement composite, **c** and **d** HF reinforced nanocomposite containing 1 wt% nanoclay, **e** and **f** HF reinforced nanocomposite containing 3 wt% nanoclay [1]

rupture and matrix fracture can be clearly seen. The examination of fracture surface of HF reinforced nanocomposites containing 1 wt% nanoclay shows good penetration of matrix between hemp filaments (Fig. 3.11c) as well as rough hemp fibre surface (Fig. 3.11d). However, poor adhesion between fibres and matrix is observed in HF reinforced cement composite (Fig. 3.11a, b). In HF reinforced nanocomposite containing 3 wt% nanoclay, macro-crack is observed which revealed relativity weak matrix as shown in Fig. 3.11e, and also debonding of fibre was occurred (Fig. 3.11f).

Figure 3.12 shows the comparison of Ca^{2+} obtained from EDS analysis on hemp fabric surfaces in different samples. EDS analysis supports that the nanoclay effectively prevented the hemp fabric degradation through pozzolanic reaction as previously described and reduces the $Ca(OH)_2$. In previous study, Sedan et al. [16] studied a deleterious effect of Ca^{2+} element on hemp fibre degradation. They reported that hemp fibre surface can interact with Ca^{2+} ions to form $Ca(OH)_2$ nodules on it, and this lead to decrease the durability of hemp fibre in alkaline environment. Figure 3.12a shows the EDS analysis of hemp fibre surface in cement paste, ratio of Ca^{2+} ions at 0.25 and 37 keV is considered high, this reveal that hemp fibres trapped and fixed Ca^{++} ions on their surfaces. On contrast, in HF reinforced nanocomposite contain 1 wt%, Fig. 3.12b, these ratios are significantly decreased. These results confirm that durability of hemp fibre increased by good consumption of Ca^{2+} ions. On the other hand, HF reinforced nanocomposite containing 3 wt% nanoclay behaves very similar to that of HF reinforced cement composite. In Fig. 3.12c, it is clearly seen that the ratio of Ca^{2+} ions at 0.25 and 37 keV again increased which leads to hemp fibre degradation. This could be attributed to the agglomeration of nanoclay at high content which leads to poor consumption of Ca^{2+} ions.

3.2.8 Thermal Stability

The thermal stability of samples was determined using thermogravimetric analysis (TGA). In this test, the thermal stability was studied in terms of the weight loss as a function of temperature in Argon atmosphere. The thermograms (TGA) of nanoclay, hemp fabric, cement paste, HF-reinforced cement composite and of HF-reinforced nanocomposites are shown in Fig. 3.13. The char yields at different temperatures are summarized in Table 3.2. For hemp fabric, it can be seen from TGA curve that the weight loss (%) between 285 and 375 °C is due to decomposition of cellulose. This result is in agreement with Rachini et al. [17] where the weight loss (%) of hemp fibres under Argon is in the range of 280–380 °C is due to cellulose decomposition. Concerning nanoclay, it can be seen from TGA curve that the weight loss (%) between 300 and 400 °C is due to decomposition of the ammonium salts on montmorillonite.

The TGA analysis show three distinct stages of decomposition in cement paste, HF-reinforced cement composite and HF-reinforced nanocomposites. The first stage of decomposition is between room temperature and 230 °C, which may be related to the decomposition of Ettringite and dehydration of C-S-H gel. The second stage of decomposition is between 420 and 500 °C, which corresponds to $Ca(OH)_2$ decomposition. The last stage of decomposition is between 670 and 780 °C, which correspond to $CaCO_3$ decomposition [18, 19]. In the first stage, HF-reinforced nanocomposites show slightly better thermal stability than cement paste due to

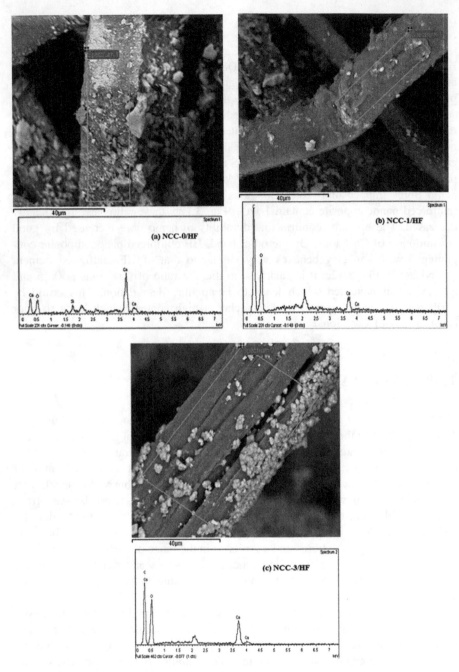

Fig. 3.12 EDS analysis with SEM images of hemp fabric surface: **a** HF reinforced cement composite, **b** HF reinforced nanocomposite containing 1 wt% nanoclay, **c** HF reinforced nanocomposite containing 3 wt% nanoclay [1]

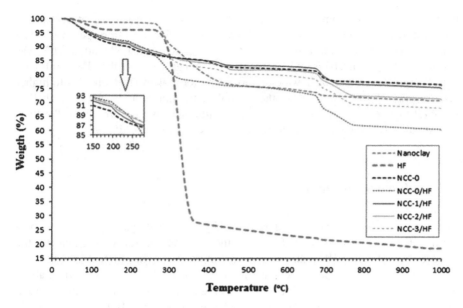

Fig. 3.13 TGA curves of nanoclay, hemp fabric (HF), cement paste, HF-reinforced cement composite and HF-reinforced nanocomposites [1]

Table 3.2 Thermal properties of nanoclay, hemp fabric (HF), cement paste, HF-reinforced cement composite and HF-reinforced nanocomposites [1]

Sample	Char yield at different temperature (%)									
	100 °C	200 °C	300 °C	400 °C	500 °C	600 °C	700 °C	800 °C	900 °C	1000 °C
Nanoclay	98.76	98.52	90.81	79.02	75.64	74.75	72.35	71.68	71.01	70.64
HF	97.01	95.85	86.89	26.79	24.75	23.06	21.40	20.48	19.36	18.48
NCC-0	93.63	89.38	86.14	84.35	82.21	81.77	78.19	77.25	76.81	76.08
NCC-0/HF	94.61	91.41	81.02	76.85	75.58	74.21	67.08	61.69	61.02	59.99
NCC-1/HF	94.27	90.32	86.23	84.70	83.03	82.76	78.46	76.30	75.81	74.85
NCC-2/HF	94.77	90.80	86.52	83.82	81.47	81.34	77.15	72.18	71.76	70.86
NCC-3/HF	94.89	91.14	85.93	82.10	79.94	79.43	74.39	69.04	68.55	67.62

resistance of nanoclay to the decomposition. In second stage, the HF-reinforced nanocomposites containing 1 wt% show better thermal stability than all samples due to dense and compact nanomatrix through consumption of CH and formation of secondary CSH gels during pozzolanic reaction [2]. Whereas, HF-reinforced nanocomposites containing 3 wt% show lower thermal stability than all cement paste and other HF-reinforced nanocomposites, in which this result confirms that slightly poor pozzolanic reaction has occurred and hence nanomatrix is less compacted. Moreover, HF-reinforced cement composites and pure nanoclay show lower thermal stability than others. At 800–1000 °C, HF-reinforced nanocomposites containing 1 wt% show thermal stability slightly less than cement paste but better than other samples. From Table 3.2 at 1000 °C, the char residue of cement paste, HF-reinforced cement composite was about 76.1 and 59.99 wt%, respectively. The char residue of HF-reinforced nanocomposites containing 1, 2 and 3 wt% was about 74.8, 70.9 and 67.6 wt%, respectively. It can be seen that HF-reinforced nanocomposites containing 1 wt% performed better in thermal stability with higher char residue of 74.85 wt% than other samples. In similar study, Chen et al. [20] reported that addition of 10 wt% nano-TiO_2 into cement paste improved the thermal stability of nanocomposite, in which it was non-reactive filler.

3.2.9   Impact Strength

The impact strength can be defined as the ability of the material to withstand impact loading [21, 22]. As shown in Fig. 3.14 the presence of nanoclay enhanced the impact strength for HF-reinforced nanocomposites. The impact strength of HF-reinforced nanocomposites containing 1 wt% nanoclay was 2.45 kJ/m^2, about 23% increase compared to HF-reinforced cement composite. This is due to good interfacial bonding between the fibres and the nanomatrix. But as clay loading increased, the impact strength decreased. For example, the impact strength of HF-reinforced nanocomposites containing 3 wt% nanoclay was 2.25 kJ/m^2, about 13% increase compared to HF-reinforced cement composite. This reduction in impact strength at higher clay loading was due to the formation of clay agglomerates and voids which led to reduced fibre-nanomatrix adhesion. Alhuthali and

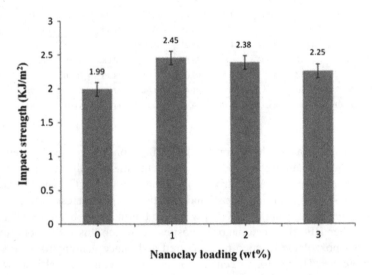

Fig. 3.14 Impact strength as a function of nanoclay content for HF reinforced composite and HF-reinforced nanocomposites [1]

Low [23] reported that the addition of 3 wt% nanoclay into recycled cellulose fibres (RCF)/vinyl ester matrix increased the impact strength by 27% compared to RCF-reinforced vinyl ester composites.

3.3 Conclusions

Cement eco-nanocomposites reinforced with hemp fabric (HF) and nanoclay platelets (Cloisite30B) are fabricated and investigated in terms of TEM, XRD, SEM, physical, thermal and mechanical properties. Results indicated that the mechanical properties generally increased as a result of the addition of nanoclay into the cement matrix with and without HF. An optimum replacement of ordinary Portland cement by 1 wt% nanoclay is concluded from the current work. It is found that, 1 wt% nanoclay decreases the porosity and also significantly increases the density, flexural strength and fracture toughness of cement composite and HF reinforced nanocomposite. The microstructural analysis results indicate that the nanoclay behaves not only as a filler to improve microstructure, but also as an activator to promote pozzolanic reaction and it prevents hemp fabric surface from degradation in high alkaline environment in cement composites. The failure micromechanisms and energy dissipative processes in HF reinforced cement composite and nanocomposite are discussed in terms of microstructural observations. In addition, the incorporation of 1 wt% nanoclay into the HF-reinforced nanocomposites improved the thermal stability and impact strength when compared to the HF-reinforced cement composite. However, the addition of more nanoclay (>1 wt%) into the HF-reinforced cement composites adversely affected the thermal, physical and mechanical properties.

References

1. Hakamy A. Microstructural design of high-performance natural fibre nanoclay cement nanocomposites. Ph.D. Thesis, Curtin University, Perth Australia; 2016.
2. Wei Y, Yao W, Xing X, Wu M. Quantitative evaluation of hydrated cement modified by silica fume using QXRD, Al MAS NMR, TG–DSC and selective dissolution techniques. Constr Build Mater. 2012;36:925–32.
3. Jo B, Kim C, Tae G, Park J. Characteristics of cement mortar with nano-SiO_2 particles. Constr Build Mater. 2007;21:1351–5.
4. Shebl S, Allie L, Morsy M, Aglan H. Mechanical behavior of activated nano silicate filled cement binders. J Mater Sci. 2009;44:1600–6.
5. Morsy MS, Alsayed SH, Aqel M. Hybrid effect of carbon nanotube and nano-clay on physico-mechanical properties of cement mortar. Constr Build Mater. 2011;25:145–9.
6. Chang T, Shih J, Yang K, Hsiao T. Material properties of Portland cement paste with nano-montmorillonite. J Mater Sci. 2007;42:7478–87.
7. Li H, Xiao H, Yuan J, Ou J. Microstructure of cement mortar with nano-particles. Compos B. 2004;35:185–9.

8. Snoeck D, De Belie N. Mechanical and self-healing properties of cementitious composites reinforced with flax and cottonised flax, and compared with polyvinyl alcohol fibres. Biosyst Eng. 2012;111:325–35.

9. De Gutiérrez R, Díaz L, Delvasto S. Effect of pozzolans on the performance of fiber-reinforced mortars. Cem Concr Compos. 2005;27:593–8.

10. Elsaid A, Dawood M, Seracino R, Bobko C. Mechanical properties of kenaf fiber reinforced concrete. Constr Build Mater. 2011;25:1991–2001.

11. Ahmed SFU, Mihashi H. Strain hardening behavior of lightweight hybrid polyvinyl alcohol (PVA) fiber reinforced cement composites. Mater Struct. 2011;44:1179–91.

12. Alamri H, Low IM. Characterization of epoxy hybrid composites filled with cellulose fibres and nano-SiC. J Appl Polym Sci. 2012;126:222–32.

13. Ahmed SFU, Maalej M, Paramasivam P. Flexural responses of hybrid steel-polyethylene fiber reinforced cement composites containing high volume fly ash. Constr Build Mater. 2007;21:1088–97.

14. Banthia N, Shengb J. Fracture toughness of micro-fiber reinforced cement composites. Cem Concr Compos. 1996;18:251–69.

15. Alamri H, Low IM. Microstructural, mechanical, and thermal characteristics of recycled cellulose fiber-halloysite-epoxy hybrid nanocomposites. Polym Compos. 2012;33:589–600.

16. Sedan D, Pagnoux C, Chotard T, Smith A, Lejolly D, Gloaguen V. Effect of calcium rich and alkaline solutions on the chemical behaviour of hemp fibres. J Mater Sci. 2007;42:9336–42.

17. Rachini A, Le Troedec M, Peyratout C, Smith A. Comparison of the thermal degradation of natural, alkali-treated and silane-treated hemp fibers under air and an inert atmosphere. J Appl Polym Sci. 2009;112:226–34.

18. Lothenbach B, Winnefeld F, Alder C, Wieland E, Lunk P. Effect of temperature on the pore solution, microstructure and hydration products of Portland cement pastes. Cem Concr Res. 2007;37:483–91.

19. Djaknoun S, Ouedraogo E, Benyahia A. Characterisation of the behaviour of high performance mortar subjected to high temperatures. Constr Build Mater. 2012;28:176–86.

20. Chen J, Kou S, Poon C. Hydration and properties of nano-TiO_2 blended cement composites. Cem Concr Compos. 2012;34:642–9.

21. Toutanji H, Xu B, Gilbert J, Lavin T. Properties of poly (vinyl alcohol) fiber reinforced high-performance organic aggregate cementitious material: converting brittle to plastic. Constr Build Mater. 2010;24:1–10.

22. Zhou X, Ghaffar S, Dong W, Oladiran O, Fan M. Fracture and impact properties of short discrete jute fibre-reinforced cementitious composites. Mater Des. 2013;49:35–47.

23. Athuthali A, Low IM, Dong C. Characterization of the water absorption, mechanical and thermal properties of recycled cellulose fibre reinforced vinyl-ester eco-nanocomposites. Compos B. 2012;43:2772–81.

Chapter 4
Nanoclay and Calcined Nanoclay-Cement Matrix: Microstructres, Physical, Mechanical and Thermal Properties

4.1 Introduction

In the construction industry, several types of nanomaterials have been incorporated into concretes or cement based materials such as nano-SiO_2, nano-Al_2O_3, nano-Fe_2O_3, nano-ZnO_2, nano-MgO, nano-$CaCO_3$, nano-TiO_2, carbon nanotubes, nano-metakaolin and nano-ZrO_2 in order to improve the durability and mechanical properties of concrete and Portland cement matrix [1–5]. Supit and Shaikh [6] reported that the addition of 1% nano-$CaCO_3$ increased the compressive strength of mortar and concrete significantly. Nano-silica (NS) has recently been introduced as an advanced pozzolan to improve the microstructure and stability of cement based system [7]. It has been observed that the NS consumed free lime (calcium hydroxide) during cement hydration and formed calcium silicate hydrate (CSH) gel due to its high fineness and reactivity [8]. In addition, the NS is particularly beneficial in acting as a nucleus to make the cement hydrate dense and improves the interfacial transition zone despite of small amount of replacement. From some conducted experiments, Zhang and Islam [9] and Jo et al. [10] reported better performance of concrete containing NS than that containing silica fume. Nanosilica also improved the microstructure and mechanical properties of high calcium fly ash based geopolymer cured at ambient temperature [11].

Nanoclay is a new generation of processed clay for a wide range of high-performance cement nanocomposite [12, 13]. As a kind of nano-pozzolanic material, nanoclay not only reduces the pore size and porosity of the cement matrix, but also improves the strength of cement matrix [14]. Furthermore, nanoclay particles enhance hardened properties of cement paste and mortar. Farzadnia et al. [15] reported that incorporation of 3% halloysite nanoclay into cement mortars increased the 28th day compressive strength up to 24% compared to the control samples. However, little research is reported on the use of calcined nanoclay as reinforcement in cement nanocomposite. In this paper, the effect of different amounts of

© The Author(s) 2017
I.-M. Low et al., *High Performance Natural Fiber-Nanoclay Reinforced Cement Nanocomposites*, Biobased Polymers, DOI 10.1007/978-3-319-56588-0_4

nanoclay and calcined nanoclay on the mechanical and thermal properties of cement nanocomposite is studied. Due to calcination the amorphous contents of nanoclay is increased, which later reacted with $Ca(OH)_2$ of the cement hydration products and formed additional calcium-silica-hydrate gel. The benefit of the use of nanoclay is the improvement of microstructure of the cement nanocomposite. This chapter presents the effect of nanoclay (NC) and calcined nanoclay (CNC) on microstructure, physical, mechanical and thermal properties of modified matrix. In this study seven series of mixes are considered. The first series is control which consisted of 100% ordinary Portland cement (OPC) and is termed a "C". The second, third and fourth series are similar to the first series in every aspect except the partial replacement of OPC by 1, 2 and 3 wt% of NC, respectively. These mixes are termed as NCC1, NCC2 and NCC3, respectively. In fifth, sixth and seventh series the effect of 1, 2 and 3 wt% of CNC as partial replacement of OPC in the first series is studied and are termed as CNCC1, CNCC2 and CNCC3, respectively. A fixed water/binder ratio of 0.485 is used in all mixes. For each series, three prismatic plate specimens of 300 mm × 70 mm × 10 mm were cast then after 56 days several rectangular specimens of each series with dimensions 70 mm × 20 mm × 10 mm were cut from the fully cured prismatic plate for mechanical and physical tests.

4.2 Results and Discussion

4.2.1 XRD Analysis

Figure 4.1 shows the XRD patterns of nanoclay and those calcined at 800, 850 and 900 °C for 2 h, respectively. XRD patterns of nanoclay show wide diffraction peaks which refer to Montmorillonite-18A $[Na_{0.3}(Al,Mg)_2Si_4O_{10}OH_2 \cdot 6H_2O]$ (PDF000120219) and also exhibit crystalline phase at 2θ of 4.82° which indicate the presence of the ammonium salt. Patterns of calcined nanoclay at 800, 850 and 900 °C show that the ammonium salt peak completely disappeared at these temperatures. Also other nanoclay peaks gradually disappeared and transformed to amorphous state (calcined nanoclay) at 900 °C (see Fig. 4.1d). Results of XRD clearly show the transformation of crystalline phases of nanoclay to amorphous phases due to calcination [16].

4.2.2 Energy Dispersive Spectroscopy Analysis

Figure 4.2a, b show typical EDS spectra of nanoclay and calcined nanoclay (at 900 °C). In Fig. 4.2a, ammonium salt in the nanoclay is identified by carbon and nitrogen elements. The content of nitrogen element is very small, thus EDS cannot

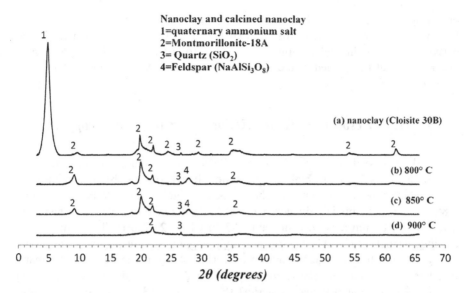

Fig. 4.1 X-ray diffraction patterns of nanoclay and calcined nanoclay [17]

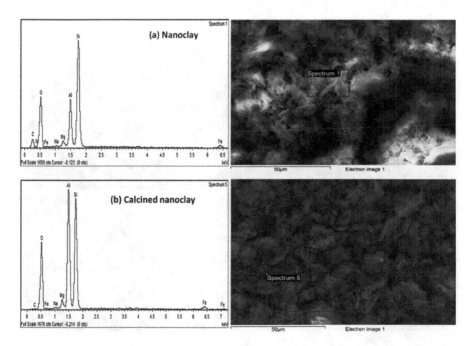

Fig. 4.2 EDS analysis with SEM images of: **a** nanoclay, **b** calcined nanoclay [17]

detect it but the carbon element is clearly detected at 2.5 keV. However, in Fig. 4.2b, the carbon element disappeared because of combustion which yielded carbon dioxide during calcination. This result also confirms the decomposition of ammonium salt in calcined nanoclay which agrees with XRD results.

4.2.3   High Resolution Transmission Electron Microscopy

HRTEM images of nanoclay (Cloisite 30B) at low and high magnification are shown in Fig. 4.3. In the high magnification image (Fig. 4.3b), it can be seen clearly that the distances between the nanoclay platelets (i.e. layers) were about 1.85 nm and thus this is evidence that the d-spacing of (001) planes in nanoclay layers was 1.85 nm as shown in Table 2.3 in Chap. 2. The HRTEM images for calcined nanoclay (at 900 °C) at low and high magnification respectively are shown in Fig. 4.3c–d. In high magnification image (Fig. 4.3d), it can be seen that many platelets in calcined nanoclay were destroyed and some of them broke to small nanoparticles with approximate spherical shapes ranging 3–8 nm. This result also confirms the amorphous phase of calcined nanoclay which agrees with the XRD results above.

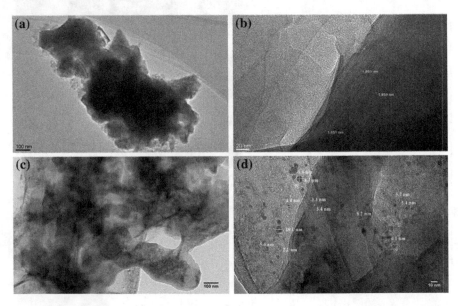

Fig. 4.3 TEM images of nanoclay and calcined nanoclay (at 900 °C) at: **a, c** low magnification, **b, d** high magnification [17]

4.3   Quantitative X-Ray Diffraction Analysis

The XRD patterns of cement paste, cement nanocomposite containing 1, 2 and 3 wt % CNC and cement nanocomposite containing 1 wt% NC are shown in Fig. 4.4a–e, that included Corundum [Al_2O_3] (PDF 000461212) phase as the internal standard. Table 4.1 shows the results of quantitative analysis with Rietveld refinement of cement paste and cement nanocomposite containing NC and CNC. Three important phases are noticed in this study: portlandite [$Ca(OH)_2$] (PDF 00-044-1481), tricalcium silicate [C_3S] (00-049-0442) and dicalcium silicate [C_2S] (PDF 00-033-0302). Moreover, four less important phases are also noticed: Ettringite [$Ca_6Al_2(SO_4)_3(OH)_{12}.26H_2O$] (PDF 000411451), Gypsum [$Ca(SO_4)(H_2O)_2$] (PDF 040154421), Quartz [SiO_2](PDF 000461045) and Calcite [$CaCO_3$](PDF 000050586) [2, 21, 23, 27]. As can be seen in Table 4.1 and Fig. 4.4b, the addition of 1 wt% CNC reduced the amount of $Ca(OH)_2$ from 16.8 to 12.1 wt%, about 28% reduction compared to cement paste. Also the intensities of major peaks of $Ca(OH)_2$

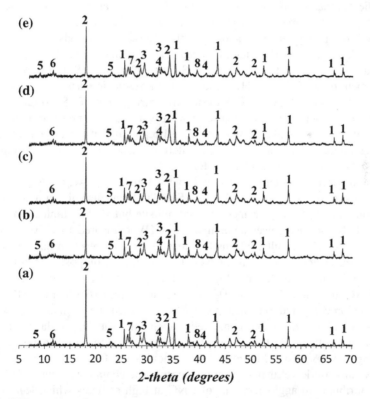

Fig. 4.4 XRD patterns of: **a** cement paste, cement nanocomposite containing: **b** 1 wt% CNC (CNCC1), **c** 2 wt% CNC (CNCC2), **d** 3 wt% CNC (CNCC3), **e** 1 wt% NC (NCC1). [*Legend* 1 = Corundum [Al_2O_3], 2 = Portlandite [$Ca(OH)_2$], 3 = Tricalcium silicate [C_3S], 4 = Dicalcium silicate [C_2S], 5 = Ettringite, 6 = Gypsum, 7 = Quartz, 8 = Calcite] [17]

Table 4.1 QXDA results for cement paste (C) and cement nanocomposite containing 1, 2 and 3 wt% CNC and 1 wt% NC [17]

Phase	Weight % (Phase abundance)				
	C	CNCC1	CNCC2	CNCC3	NCC1
Portlandite [$Ca(OH)_2$]	16.8	12.1	13.2	14.1	13.8
Ettringite [$Ca_6Al_2(SO_4)_3(OH)_{12}.26H_2O$]	2.0	1.3	1.5	1.8	1.6
Tricalcium silicate [C_3S]	1.3	2.0	1.7	1.4	1.5
Dicalcium silicate [C_2S]	4.4	6.6	6.1	5.4	6.1
Gypsum [$Ca(SO_4)(H_2O)_2$]	0.7	0.4	0.6	0.4	0.4
Calcite [$CaCO_3$]	3.7	2.1	2.7	3.3	3.0
Quartz [SiO_2]	0.9	0.6	0.4	0.7	0.5
Amorphous content	70.1	74.8	73.7	72.8	73.0

were significantly reduced compared to cement paste (Fig. 4.4a–b). Furthermore, the amorphous content was increased from 70.1 to 74.8 wt%, about 6.7% increase. This indicates that an obvious consumption of $Ca(OH)_2$ crystals mainly due to the effect of pozzolanic reaction in the presence of amorphous CNC and good dispersion of amorphous calcined nanoclay in the matrix, which leads to more calcium silicate hydrate gel (C-S-H) being produced. This explanation is also confirmed by an increase in the amount of unreacted C_3S (2.0 wt%) and C_2S (6.6 wt%), relative to the cement paste. Wei et al. [18] reported that pozzolanic reaction decelerates the hydration reaction of C_3S and C_2S during the curing time of 28–90 days. In this study, these unreacted phases could react with water later to produce more C-S-H gel after 56 days. Recently, Shaikh et al. [19] reported that the cement nanocomposite containing 2 wt% nano-silica exhibited less calcium hydroxide but slightly more C_2S than the control cement paste.

On the other hand, as can be seen in Table 4.1 and Fig. 4.4e, the NCC1 cement nanocomposite shows lower amounts of C_3S and C_2S and also higher amount of Ca $(OH)_2$ compared to CNCC1 cement nanocomposite but slightly higher amounts of C_3S and C_2S and also lower amount of $Ca(OH)_2$ compared to CNCC3 cement nanocomposite. This result confirms that less pozzolanic reaction has occurred in NCC1 cement nanocomposite than CNCC1 cement nanocomposite. In contrast, as can be seen from Table 4.1 and Fig. 4.4d for cement nanocomposite containing 3 wt% CNC, the amount of $Ca(OH)_2$ was decreased from 16.8 to 14.1 wt%, about 16% reduction compared to cement paste. Also the intensities of major peaks of Ca $(OH)_2$ were slightly decreased compared to cement paste (Fig. 4.4a, d). But this reduction of amount of $Ca(OH)_2$ is less than the reduction in cement nanocomposite containing 1 wt% CNC. Moreover, the amounts of C_3S (1.4 wt%) and C_2S (5.4 wt%) are also lower than cement nanocomposite containing 1 wt% CNC. This may be attributed to agglomerations of CNC at high contents which lead to relatively poor dispersion of CNC and hence relatively poor pozzolanic reaction [20]. Table 4.1 also shows that the calcite content varies in all samples. For example, the content of calcite decreased from 3.7 to 2.1 wt% in cement nanocomposite containing 1 wt% CNC. This indicates that little carbonation occurred over the 56 day curing period. Table 4.1 shows that Ettringite is slightly less in cement nanocomposite than cement paste. For example, it decreased from 2.0 to 1.3 wt% in cement nanocomposite containing 1 wt% CNC.

4.3.1 Calculation of Ca(OH)₂ Content in Cement Nanocomposites

The Ca(OH)$_2$ content (CH) is calculated from the TGA curves using the following equation [21]:

$$CH(\%) = WL_{CH}(\%) \frac{MW_{CH}}{MW_{H_2O}} \tag{4.1}$$

where $WL_{CH}(\%)$ is corresponds to the weight loss attributable to Ca(OH)$_2$ decomposition, MW_{CH} is the molecular weight of CH (74.01 g/mol) and MW_{H_2O} is the molecular weight of H$_2$O (18 g/mol).

The thermograms (TGA) of cement paste and cement nanocomposite containing CNC and NC are shown in Fig. 4.5. Table 4.2 summarises the CH content of above measured by QXDA and TGA techniques. Results in Table 4.2 indicate that TGA is at least as good as QXDA for quantifying the amount of calcium hydroxide [22]. It can be seen that there is good agreement between the two techniques, where both measured amounts are very close to each other [28]. However, the amounts of CH by TGA are slightly lower than the QXDA. This observation is in agreement with the work done by Scrivenera et al. [23] and Korpa et al. [24], in which they reported that this discrepancy could be attributed to the possible error sources of each method itself that were difficult to quantify. Nevertheless, inside the error margin there is a

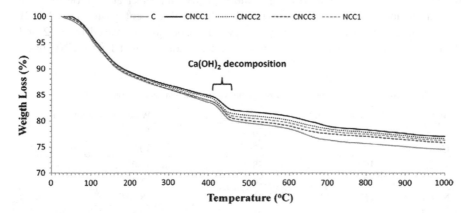

Fig. 4.5 TGA curves of cement paste (C) and cement nanocomposite: CNCC1, CNCC2, CNCC3 and NCC1 [17]

Table 4.2 Calculation of Ca(OH)$_2$ content in cement paste and cement nanocomposite containing 1, 2 and 3 wt% CNC and 1 wt% NC by QXDA and TGA techniques. [17]

Sample	TGA (wt%)	QXRD (wt%)	Difference (wt%)
C	15.5	16.8	1.3
CNCC1	10.7	12.1	1.4
CNCC2	12.1	13.2	1.1
CNCC3	13.0	14.1	1.1
NCC1	12.3	13.8	1.5

good correlation of the values assessed by both techniques employed [24]. And also the above consistency added a new evidence for reliability of the QXRD method to characterize quantitatively the hydration of cement systems [28]. The TGA and QXDA results in Table 4.2 also confirm the reactivity of 1 wt% CNC in reducing the CH content in cement nanocomposite. The CNC is mainly amorphous material and behaves as a highly reactive artificial pozzolan. The CH content by the TGA and QXDA in cement nanocomposite containing 1 wt% CNC was 10.7 and 12.1 wt%, respectively. It is also be seen that the CH content in cement nanocomposite containing 1 wt% CNC is reduced significantly when compared to cement paste and cement nanocomposite containing NC and CNC such as cement nanocomposite containing 1 wt% NC. This could be due to the reactivity of 1 wt% CNC in cement nanocomposite and the consumption of CH by the pozzolanic reaction.

4.3.2 Porosity, Water Absorption and Density

The porosity, water absorption and density of cement paste and cement nanocomposite containing NC and CNC are shown in Table 4.3. It is noticed that the addition of CNC or NC decreases the porosity and water absorption of these cement nanocomposites when compared to control cement paste. In CNCC1 cement nanocomposite, the porosity and water absorption decreased by 31.2 and 34%, respectively compared to cement paste. This indicates that 1 wt% CNC has a

Table 4.3 Porosity, density and water absorption values for various samples [17]

Sample	Porosity (%)	Density (g/cm^3)	Water absorption (%)
C	23.9	1.76	13.4
NCC1	18.7	1.87	10.2
NCC2	19.6	1.78	11.0
NCC3	19.9	1.76	11.3
CNCC1	16.5	1.93	8.9
CNCC2	17.6	1.91	9.6
CNCC3	18.9	1.85	10.3

filling effect in the porosity of cement nanocomposite. This result is in agreement with the work done by Jo et al. [10] where the porosity of cement mortar is decreased by the addition of nano-SiO_2 particles. Supit and Shaikh [25] reported that the addition of 2 wt% nano-silica significantly reduced the porosity of high volume fly ash (HVFA) concrete. Furthermore, in Table 4.3, the addition of 1 wt% CNC increased the density of control cement paste from 1.76 to 1.93 g/cm^3, about 9.7% increase. This improvement demonstrates that cement nanocomposite with 1 wt% CNC yields consolidated denser microstructure. However, further addition of CNC or NC leads to an increase in porosity and water absorption and a decrease in density. This could be attributed to the poor dispersion and agglomerations of the high CNC or NC contents which create more voids in the matrix [26].

SEM examinations of the microstructure of cement paste, CNCC1and CNCC3 cement nanocomposite are shown in Fig. 4.6. For cement paste, Fig. 4.6a shows more Ca(OH)$_2$ crystals and Ettringite as well as more pores which revealed a weak structure. Figure 4.6b shows the SEM micrograph of CNCC1, which is different from that of cement paste, the structure is dense and compact with few pores and

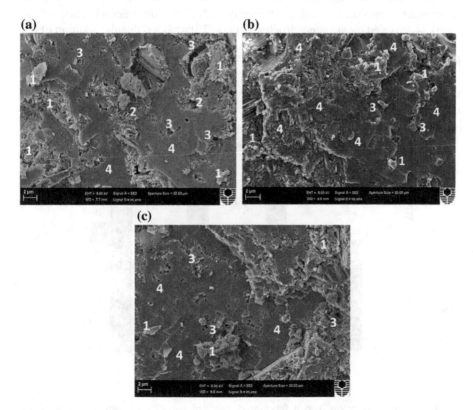

Fig. 4.6 SEM micrographs of: **a** cement paste, cement nanocomposite containing: **b** 1 wt% CNC, **c** 3 wt% CNC. (*Legend* 1 = [Ca(OH)$_2$], 2 = Ettringite, 3 = pores, 4 = C-S-H gel) [17]

more C-S-H gel. On the other hand, in Fig. 4.6c, the CNCC3 shows more pores than CNCC1which relatively weaken the structure.

4.3.3 Compressive Strength

The compressive strength of the cement paste, cement nanocomposite containing NCC and CNCC are presented in Fig. 4.7. It can be noticed from results in Fig. 4.7 that the addition of NC and CNC to cement paste increases the compressive strength of all cement nanocomposite pastes. For instance, the cement nanocomposite containing 1 wt% CNC exhibited an enhancement in the compressive strength from 53.1 to 74.2 MPa or 40% increase, whereas in the cement nanocomposite containing 1 wt% NC, the compressive strength reached 69.8 MPa. The increase in compressive strength of cement nanocomposite containing 1 wt% CNC is due to amorphous state of CNC (i.e. small particle size) and extremely large surface area, in which the CNC reacts more quickly with free lime in the hydration reaction than NC and subsequently produced more secondary C–S–H gel and filled the capillary pores in the matrix efficiently [27]. Thus the microstructure of the matrix is densified by the nanoparticles. Chang [28] reported that the addition of 0.6 wt% nano-montmorillonite into cement paste increased compressive at age of 56 days from 46 to 52.1 MPa (i.e. 13.2% increases) compared to the cement paste. Li et al. [29] noticed 26% improvement in 28 days compressive strength of cement mortar containing 3% nano silica. Despite benefits of CNC and NC, it is important to note that the nano particles have a tendency to agglomerate when using at high content (i.e. more 3 wt% CNC) in the mixes [30]. This aggregation forms

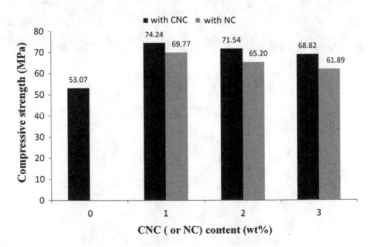

Fig. 4.7 Compressive strength as a function of calcined nanoclay (or nanoclay) content for cement paste and cement nanocomposite [17]

weak zones and consequently prevents the formation of homogenous hydrate microstructure. Therefore, the appropriate proportion of CNC content should be taken into account.

4.3.4   Flexural Strength

Flexural strengths of cement paste, cement nanocomposite containing NC and CNC are shown in Fig. 4.8. Overall, the incorporation of CNC or NC into the cement matrix led to significant enhancement in the flexural strength of all cement nanocomposites. The flexural strength of cement nanocomposite containing 1, 2 and 3 wt% CNC is increased by 42.9, 34.8 and 30.6%, respectively compared to cement paste. While the flexural strength of cement nanocomposite containing 1, 2 and 3 wt% NC is increased by 32.1, 29.3 and 24.7% respectively compared to cement paste. This improvement clearly indicates the effectiveness of CNC in consuming calcium hydroxide (CH), supporting pozzolanic reaction and filling the micro pores in the matrix [30]. Thus the microstructure of cement nanocomposite is denser than the cement matrix, especially in the case of using 1 wt% CNC, which is evident from its higher flexural strength. Hosseini et al. [31] studied the effect of nanoclay (Cloisite15A) on the mechanical properties of cement mortar at 28 days with water/binder ratio of 0.4. They reported that addition of 1 wt% nanoclay improved the flexural strength from 7.0 to 9.1 MPa, about 30% increase. Qing et al. [8] studied the influence of 3 wt% nano-SiO$_2$ (NS) addition on properties of hardened cement paste. They observed that the flexural strength increased by about 72% compared to control cement matrix. They attributed this improvement to the pozzolanic and filler effects of nano-SiO$_2$ particles.

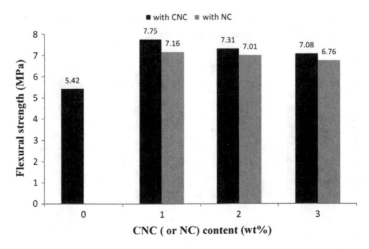

Fig. 4.8 Flexural strength as a function of calcined nanoclay (or nanoclay) content for cement paste and cement nanocomposite [17]

However, the addition of more than 1 wt% CNC caused a marked reduction in the flexural strength. This could be attributed to the relatively poor dispersion and agglomerations of the CNC in the cement matrix at higher CNC contents, which create weak zones, in the form of micro-voids which cause stress concentration [32]. Moreover, the addition of more CNC (i.e. 2 wt%) led to a significant reduction in the flexural strength due to an increase in porosity. Nevertheless the addition of CNC improved the flexural strength of cement nanocomposite. For example, in this study, although the flexural strength of cement nanocomposite with 3 wt% CNC decreased compared to cement nanocomposite with 1 wt% CNC but it is still higher than the control cement paste.

4.3.5 Fracture Toughness

Fracture toughness of cement paste and cement nanocomposite containing NCC and CNCC are shown in Fig. 4.9. The fracture toughness of cement nanocomposite containing 1, 2 and 3 wt% CNC were 0.49, 0.47 and 0.44 MPa.m$^{1/2}$, respectively. It can be seen that the fracture toughness of CNCC1 cement nanocomposite is increased by 40% compared to cement paste. This is attributed to the fact that the CNC modified the matrix through pozzolanic reaction and reduced the Ca(OH)$_2$ content. Alamri and Low [33] reported that the addition of 1 wt% halloysite nanotubes (HNTs) into epoxy matrix significantly increased the fracture toughness from 0.85 to 1.33 MPa.m$^{1/2}$ (i.e. by 56.5%) compared to epoxy matrix. However, facture toughness of CNCC cement nanocomposite gradually decreased when CNC contents are increased after the optimum content of 1 wt%. This is attributed to the poor dispersion of high content of CNC into the matrix, which leads to increase in porosity [34].

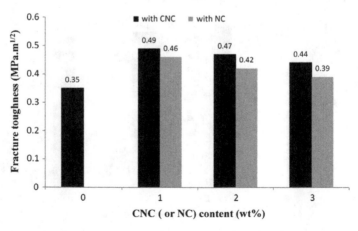

Fig. 4.9 Fracture toughness as a function of calcined nanoclay (or nanoclay) content for cement paste and cement nanocomposite [17]

4.3.6 Impact Strength

The impact strength is defined as the ability of the material to withstand impact loading [35]. The impact strengths of cement paste and cement nanocomposite containing NCC and CNCC are shown in Table 4.4. Generally, it can be seen that the impact strength of cement paste is significantly improved due to the addition of CNC or NC. The impact strength of NCC1 cement nanocomposite is 3.1 kJ/m^2, about 29.4% increase compared to the cement paste. While the impact strength of CNCC1 cement nanocomposite is 3.2 kJ/m^2, about 33.6% increase compared to cement paste. Alamri and Low [36] reported that the addition of 5 wt% nanoclay to epoxy matrix increased the impact strength from 5.6 to 7.8 kJ/m^2 about 39.3% increase compared to epoxy matrix. However, as CNC loading increased after the optimum content of 1 wt% the impact strength is decreased. For example, the impact strength of CNCC3 cement nanocomposite was 3.1 kJ/m^2, about 4% decrease compared to CNCC1 cement nanocomposite. This reduction in impact strength at higher CNC loading was due to the formation of CNC agglomerates and voids which led to weaken nanocomposite.

4.3.7 Rockwell Hardness

The Rockwell hardness of cement paste and cement nanocomposite containing NCC and CNCC are shown in Table 4.4. Generally, the addition of CNC or NC into the cement matrix led to significant enhancement in the Rockwell hardness of all cement nanocomposites. As shown in Table 4.4 the Rockwell hardness of cement nanocomposite containing 1, 2 and 3wt% CNC were 91.3, 89.0 and 86.3 HRH, respectively, which corresponds to about 31.1, 27.7 and 23.9%, respectively increase compared to cement paste. While the Rockwell hardness of cement nanocomposite containing 1, 2 and 3 wt% NC is increased by 25.3, 21.0 and 18.6% respectively compared to the cement paste. This improvement demonstrates that the microstructure of cement nanocomposite is denser than the cement matrix,

	Sample	Impact strength (kJ/m^2)	Rockwell hardness (HRH)
Table 4.4 Impact strength and Rockwell hardness values for cement paste (C), (NCC) cement nanocomposite containing NC and (CNCC) cement nanocomposite containing CNC [17]	C	2.38 ± 0.06	70 ± 1
	NCC1	3.08 ± 0.15	87 ± 2
	NCC2	3.01 ± 0.11	84 ± 1
	NCC3	2.92 ± 0.08	83 ± 1
	CNCC1	3.18 ± 0.05	91 ± 1
	CNCC2	3.14 ± 0.08	89 ± 1
	CNCC3	3.05 ± 0.14	86 ± 2

Table 4.5 Thermal properties of cement paste (C) and cement nanocomposite containing 1, 2 and 3 wt% CNC and cement nanocomposite containing 1 wt% NC [17]

Sample	Char yield (%) at different temperature (°C)									
	100	200	300	400	500	600	700	800	900	1000
C	95.8	88.8	86.0	83.6	79.6	78.4	76.4	75.7	75.1	74.6
CNCC1	96.19	89.32	86.83	84.88	81.75	80.86	79.01	78.33	77.61	77.10
CNCC2	96.02	89.20	86.59	84.57	81.04	80.09	78.51	77.86	77.18	76.67
CNCC3	95.82	88.82	86.03	83.96	80.03	79.02	77.60	77.05	76.45	75.93
NCC1	95.72	89.10	86.37	83.96	80.56	79.60	78.12	77.49	76.85	76.35

especially in the case of using 1 wt% CNC. That is because of the efficiency of CNC in promoting pozzolanic reaction and filling effect [37, 38]. In an analogous research, Gupta et al. [39] reported that the hardness number (HRH) of the Fe-Al$_2$O$_3$ metal matrix nanocomposite was much higher in comparison to the cast iron specimen. However, the addition of high CNC or NC contents e.g. 3% did not show any improvement in the hardness when compared to 1 wt% CNC.

4.3.8 Thermal Stability

Weight loss (%) curves of cement paste and cement nanocomposite containing CNCC and NCC1 are shown in Fig. 4.10. The char yields at different temperatures are summarized in Table 4.5. The TGA analysis shows three distinct stages of decomposition in these samples. The first stage of decomposition is between room 230 °C, which may be related to the decomposition of Ettringite and dehydration of C-S-H gel (loss of water). The second stage of decomposition is between 400 and 510 °C, which corresponds to Ca(OH)$_2$ decomposition. The third stage of decomposition is between 670 and 780 °C, which correspond to CaCO$_3$ decomposition [40, 41]. In the first stage, generally all cement nanocomposites exhibited slightly better thermal stability than cement paste due to higher resistance of CNC or NC to decomposition [42]. Concerning the cement nanocomposite containing CNCC in second and third stage, the CNCC1 cement nanocomposite shows better thermal stability than CNCC2, CNCC3 and NCC1 cement nanocomposite due to dense and compact nanocomposite through consumption of calcium hydroxide and formation of secondary CSH gels during pozzolanic reaction [18]. In contrast, NCC1 cement nanocomposite shows lower thermal stability than CNCC2 cement nanocomposite but slightly higher than CNCC3 cement nanocomposite. This result confirms that slightly poor pozzolanic reaction has occurred and hence this NCC1 nanocomposite is less dense when compared to CNCC1 and CNCC2 cement nanocomposites. From Table 4.5 at 1000 °C, the char residue of cement paste, CNCC1, CNCC2 and CNCC3 cement nanocomposite was about 74.6, 77.1, 76.7 and 75.9 wt%, respectively. It can be seen that the CNCC1 cement nanocomposite performed better in thermal stability with higher char residue of about 3.3 and 1.5% more than cement paste and CNCC3 cement nanocomposite, respectively. In a similar study, Chen et al. [43] reported that addition of 10 wt% nano-TiO$_2$ into cement paste improved the thermal stability of cement nanocomposite considerably.

4.4 Cost-Benefit Analysis and Applications

There is a huge optimism on the use of nanomaterials in construction and building applications although the nanoparticles are expensive and could limit their applications [44, 45]. However, nano particles exhibit unique characteristics which result in new generation of concrete that is stronger and more durable [46]. With progress of manufacturing technologies the cost of nano particles is also expected to drop in future. Moreover, the nanoparticles are used in very small amount in the concrete or other cementitious nanocomposites. For example, in this study 1 wt% calcined nanoclay in cement nanocomposite led significant improvement in mechanical properties. From economic point of view, the addition of 1% calcined nanoclay in cement nanocomposite will not add any significant cost but improved the mechanical properties by about 40%. Shaikh and Supit [28] stated that although the use of nano-CaCO$_3$ was first considered as filler to partially replace cement or gypsum, some studies have shown advantages of using 1% nano-CaCO$_3$ nanoparticles in terms of compressive strength, accelerating effect and economic benefits as compared to cement and other supplementary cementitious materials.

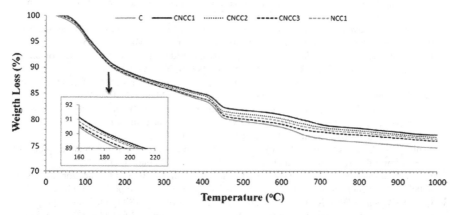

Fig. 4.10 Weight loss (%) curves by TGA of cement paste (C) and cement nanocomposite: CNCC1, CNCC2, CNCC3 and NCC1 [17]

4.5   Conclusions

The influence of nanoclay (NC) and calcined nanoclay (CNC) on the mechanical and thermal properties of cement nano-composites presented. Calcined nanoclay is prepared by heating nanoclay (Cloisite 30B) at 900 °C for 2 h. Characterisation of microstructure is investigated using Quantitative X-ray Diffraction Analysis (QXDA) and High Resolution Transmission Electron Microscopy (HRTEM). Estimation of $Ca(OH)_2$ content in the cement nanocomposite is studied by the combination of QXDA and thermogravimetry analysis (TGA) techniques. Results showed that the mechanical and thermal properties of the cement nanocomposites are improved as a result of NC and CNC addition. An optimum replacement of ordinary Portland cement with 1 wt% CNC is observed through reduced porosity and water absorption as well as increased density, compressive strength, flexural strength, fracture toughness, impact strength, hardness and thermal stability of cement nanocomposites. The microstructural analyses from QXRA and SEM indicate that the CNC acted not only as a filler to improve the microstructure, but also as the activator to support the pozzolanic reaction. Cost-benefit analysis indicates that nanoparticles are expensive but from economic point of view nanoclay is used in very small amount (i.e. 1 wt%) in cementitious materials. As a result nanoclay does not add any significant cost but improves the mechanical properties significantly.

References

1. Nazari A, Riahi S (2011) The effects of zinc oxide nanoparticles on flexural strength of self-compacting concrete. Compos B 42:167–75
2. Morsy MS, Alsayed SH, Aqel M (2011) Hybrid effect of carbon nanotube and nano-clay on physico-mechanical properties of cement mortar. Constr Build Mater 25:145–49
3. Li H, Xiao H, Guan X, Wang Z, Yu L (2014) Chloride diffusion in concrete containing nano-TiO_2 under coupled effect of scouring. Compos B 56:698–704
4. Nazari A, Riahi S (2012) The effects of ZrO_2 nanoparticles on properties of concrete using ground granulated blast furnace slag as binder. J Compos Mater 46:1079–90
5. Moradpour R, Taheri-Nassaj E, Parhizkar T, Ghodsian M (2013) The effects of nanoscale expansive agents on the mechanical properties of non-shrink cement-based composites: the influence of nano-MgO addition. Compos B 55:193–202
6. Supit S, Shaikh FUA (2014) Effect of nano-$CaCO_3$mon compressive strength development of high volume fly ash mortars and concretes. J Adv Concr Technol 12:178–86
7. Hou P, Kawashima S, Kong D, Corr D, Qian J, Shah S (2013) Modification effects of colloidal nano-SiO_2 on cement hydration and its gel property. Compos B 45:440–8
8. Qing Y, Zenan Z, Deyu K, Rongshen C (2007) Influence of Nano-SiO_2 addition on properties of hardened cement paste as compared with silica fume. Constr Build Mater 21:539–545
9. Zhang MH, Islam J (2012) Use of nano-silica to reduce setting time and increase earlystrength of concretes with high volume fly ash or slag. Constr Build Mater 29:573–80
10. Jo BW, Kim CH, Lim JH (2007) Characteristics of cement mortar with nano-silica particles. ACI Mater J 104:404–7

11. Phoo-ngernkham T, Chindaprasirt P, Sata V, Hanjitsuwan S (2014) The effect of adding nano-silica and nano-alumina on properties of high calcium fly ash geopolymer cured at ambient temperature. Mater Des 55:58–65

12. Aly M, Hashmi MSJ, Olabi AG, Messeiry M, Hussain AI (2011) Effect of nano clay particles on mechanical, thermal and physical behaviours of waste glass cement mortars. Mater Sci Eng, A 528:7991–8

13. Alamri H, Low IM, Alothman Z (2012) Mechanical, thermal and microstructural character-istics of cellulose fibre reinforced epoxy/organoclay nanocomposites. Compos B 43:2762–71

14. Wei J, Meyer C (2014) Sisal fiber-reinforced cement composite with Portland cement substitution by a combination of metakaolin and nanoclay. J Mater Sci 49:7604–19

15. Farzadnia N, Ali A, Demirboga R, Anwar M (2013) Effect of halloysite nanoclay on mechanical properties, thermal behavior and microstructure of cement mortars. Cem Concr Res 48:97–104

16. He C, Makovicky E, Osbaeck B (1996) Thermal treatment and pozzolanic activity of Na- and Ca-montmorillonite. Appl Clay Sci 10:351–68

17. Hakamy A. Microstructural design of high-performance natural fibre nanoclay cement nanocomposites. Ph.D. Thesis; 2016 Curtin University, Perth Australia

18. Wei Y, Yao W, Xing X, Wu M (2012) Quantitative evaluation of hydrated cement modified by silica fume using QXRD, 27Al MAS NMR, TG–DSC and selective dissolution techniques. Constr Build Mater 36:925–32

19. Shaikh FUA, Supit SWM, Sarker PK (2014) A study on the effect of nano silica on compressive strength of high volume fly ash mortars and concretes. Mater Des 60:433–42

20. Najigivi A, Khaloo A, Iraji A, Abdul-Rashid S (2013) Investigating the effects of using different types of SiO_2 nanoparticles on the mechanical properties of binary blended concrete. Compos B 54:52–8

21. Shaikh FUA, Supit S (2014) Mechanical and durability properties of high volume fly ash (HVFA) concrete containing calcium carbonate ($CaCO_3$) nanoparticles. Constr Build Mater 70:309–21

22. Soin A, Catalan L, Kinrade S (2013) A combined QXRD/TG method to quantify the phase composition of hydrated Portland cements. Cem Concr Res 48:17–24

23. Scrivenera K, Fullmanna T, Galluccia E, Walentab G, Bermejob E (2004) Quantitative study of Portland cement hydration by X-ray diffraction/Rietveld analysis and independent methods. Cem Concr Res 34:1541–7

24. Korpa A, Kowald T, Trettin R (2009) Phase development in normal and ultra-high performance cementitious systems by quantitative X-ray analysis and thermoanalytical methods. Cem Concr Res 39:69–76

25. Supit SWM, Shaikh FUA (2014) Durability properties of high volume fly ash concrete containing nano-silica. Mater Struct. doi:10.1617/s115270140329-0

26. Senff L, Tobaldi D, Lucas S, Hotza D, Ferreira V, Labrincha J (2013) Formulation of mortars with nano-SiO_2 and nano-TiO_2 for degradation of pollutants in buildings. Compos B 44:40–7

27. Stefanidou M, Papayianni I (2012) Influence of nano-SiO_2 on the Portland cement pastes. Compos B 43:2706–10

28. Chang T, Shih J, Yang K, Hsiao T (2007) Material properties of portland cement paste with nano-montmorillonite. J Mater Sci 42:7478–87

29. Li H, Xiao H, Yuan J, Ou J (2004) Microstructure of cement mortar with nano-particles. Compos B 35:185–9

30. Shebl S, Allie L, Morsy M, Aglan H (2009) Mechanical behavior of activated nano silicate filled cement binders. J Mater Sci 44:1600–6

31. Hosseini P, Hosseinpourpia R, Pajum A, Khodavirdi M, Izadi H, Vaezi A (2014) Effect of nano-particles and aminosilane interaction on the performances of cement-based composites: An experimental study. Constr Build Mater 66:113–24

32. Givi A, Abdul-Rashid S, Aziz F, Salleh M (2010) Experimental investigation of the size effects of SiO_2 nano-particles on the mechanical properties of binary blended concrete. Compos B 41:673–7

33. Alamri H, Low IM (2012) Microstructural, mechanical, and thermal characteristics of recycled cellulose fiber-halloysite-epoxy hybrid composites. Polym Compos 33:589–600
34. Givi A, Abdul-Rashid S, Aziz F, Salleh M (2011) The effects of lime solution on the properties of SiO_2 nanoparticles binary blended concrete. Compos B 42:562–9
35. Zhou X, Ghaffar S, Dong W, Oladiran O, Fan M (2013) Fracture and impact properties of short discrete jute-fiber reinforced cementitious composites. Mater Des 49:35–47
36. Alamri H, Low I.M. Effect of water absorption on the mechanical properties of nano-filler reinforced epoxy nanocomposites. Mater Des 2012;42:214–22.
37. Yu M, George C, Cao Y, Wootton D, Zhou J (2014) Microstructure, corrosion, and mechanical properties of compression-molded zinc-nanodiamond composites. J Mater Sci 49:3629–41
38. Karimzadeh A, Ayatollahi M (2012) Investigation of mechanical and tribological properties of bone cement by nano-indentation and nano-scratch experiments. Polym Test 31:828–33
39. Gupta P, Kumar D, Parkash O, Jha A. Sintering and Hardness Behavior of $Fe-Al_2O_3$ Metal Matrix Nanocomposites Prepared by Powder Metallurgy. *J Compos*. 2014. doi.10.1155/2014/145973
40. Lothenbach B, Winnefeld F, Alder C, Wieland E, Lunk P (2007) Effect of temperature on the pore solution, microstructure and hydration products of Portland cement pastes. Cem Concr Res 37:483–91
41. Djaknoun S, Ouedraogo E, Benyahia A (2012) Characterisation of the behaviour of high performance mortar subjected to high temperatures. Constr Build Mater 28:176–86
42. Alhuthali A, Low IM, Dong C (2012) Characterization of the water absorption, mechanical and thermal properties of recycled cellulose fibre reinforced vinyl-ester eco-nanocomposites. Compos B 43:2772–81
43. Chen J, Kou S, Poon C (2012) Hydration and properties of nano-TiO_2 blended cement composites. Cem Concr Compos 34:642–9
44. Pacheco-Torgal F, Jalali S (2011) Nanotechnology: Advantages and drawbacks in the field of construction and building materials. Constr Build Mater 25:582–90
45. Sanchez F, Sobolev K (2010) Nanotechnology in concrete—A review. Constr Build Mater 24:2060–71
46. Singh T (2014) A review of nanomaterials in civil engineering works. Inter J Struct Civ Eng Res 3:31–5

Chapter 5
Chemically-Treated Hemp Fabric and Calcined Nanoclay Reinforced Cement Nanocomposites: Microstructures, Physical, Mechanical and Thermal Properties

5.1 Introduction

Recently, natural fibres are gaining increasing popularity to develop 'environmental-friendly construction materials' as alternative to synthetic fibres in fibre-reinforced concrete [1–3]. Natural and cellulose fibres have been used in polymer and cement matrices to improve their tensile/flexural strength and fracture resistance properties [4, 5]. They are cheaper, biodegradable and lighter than synthetic fibres. Some examples of natural fibres are: sisal, flax, hemp, bamboo and others [6–8]. Some researchers have shown that pre-treatments of natural fibre surfaces either via pulping processes such as the Kraft process or some chemical agents such as alkalization, PEI (polyethylene imine), $Ca(OH)_2$ and $CaCl_2$ have slightly improved the interfacial bond strength between natural fibres and the matrix of the eco-composites. As a result, the mechanical properties of such materials are enhanced [9–11]. Troedec et al. [12] reported that the modification of hemp fibres with NaOH has improved the interfacial bonding between the fibres and the lime-based mineral matrix (mortar).

On the other hand, one of the most effective techniques to obtain a high performance cementitious composite is by reinforcement with textile fabric, which is impregnated with cement paste or mortar. Synthetic textile fabrics such as polyethylene (PE) and polypropylene (PP) have been used as reinforcement for cement composites, in which the fabrics are made of multi-filaments. When compared to continuous or short fibres, this system has superior filament-matrix bonding which improves the tensile and flexural strength [13–16]. The use of natural fibre sheets and fabrics is more prevalent in polymer matrix when compared to cement-based matrix [17].

However, the interfacial bonding between the natural fibre and the cement matrix is relatively weak and also the degradation of fibres in a high alkaline environment of cement can adversely affect the mechanical and durability properties of natural fibre reinforced cement composites [18]. However, little or no

© The Author(s) 2017
I.-M. Low et al., *High Performance Natural Fiber-Nanoclay Reinforced Cement Nanocomposites*, Biobased Polymers, DOI 10.1007/978-3-319-56588-0_5

research has reported on the combined use of calcined nanoclay (CNC) and hemp fabrics as hybrid reinforcement in cement-composites. In this chapter the effect of CNC in hemp fibre-reinforced cement composite to overcome the above-mentioned disadvantages of hemp fibres in cementitious composites is presented. The effect of CNC on the microstructural and mechanical properties of chemically-treated hemp fabric-reinforced cement composite is also presented. In this study 12 series of mixes are considered. The first series is control series consisted of 100% OPC without NC and HF. This series is termed as "C". In second to fifth series 4, 5, 6 and 7 layers of untreated FF are used to reinforce the matrix in series 1 and are termed as 4UHFRC, 5UHFRC, 6UHFRC and 7UHFRC, respectively. The composite in sixth series contained 6 layers of NaOH treated FF with 100% OPC matrix and is termed as 6THFRC. The seventh, eighth and ninth series are similar to the first series in every aspect except the OPC is partially replaced by 1, 2 and 3 wt% of CNC and are termed as CNCC1, CNCC2 and CNCC3, respectively. In tenth, eleventh and twelveth series NaOH treated 6 layer of HF are used to reinforced the CNC matrix above and are termed as 6THFRC-CNCC1, 6THFRC-CNCC2 and 6THFRC-CNCC3, respectively. Detail mix porportions are shown in Table 2.6 of Chap. 2.

5.2 Results and Discussion

5.2.1 XRD Analysis of Calcined Nanoclay

Figure 5.1a–d shows the XRD patterns of as-received nanoclay and calcined nanoclay at 800, 850 and 900 °C for 2 h, respectively. Four phases have been indexed in the diffraction pattern of nanoclay (Fig. 5.1a) with the major phase being Montmorillonite $[(Ca,Na)_{0.3}Al_2(Si,Al)_4O_{10}(OH)_2 \cdot xH_2 \ O]$ (PDF 00052039), and minor phases of Cristobalite $[SiO_2]$ (PDF 000391425), Quartz $[SiO_2]$ (PDF 000470718) and the quaternary ammonium salt (PDF 000571718). Montmorillonite has five major peaks in the XRD pattern that correspond to 2θ of 4.84°, 19.74°, 35.12°, 53.98° and 61.80°. Each of Cristobalite and Quartz has a peak that corresponds to 2θ of 21.99° and 26.61° respectively. The quaternary ammonium salt has four peaks that correspond to 2θ of 4.84°, 9.55°, 24.42° and 29.49°. Note that there was an overlap of peaks at 2θ of 4.84° for Montmorillonite and quaternary ammonium salt. However these peaks disappeared after calcination due to the decomposition of the latter in calcined nanoclay.

In Fig. 5.1b–c, the diffraction peaks of calcined nanoclay at 800 and 850 °C are related to heated-Montmorillonite $[NaMgAlSi_4O_{11}]$ (PDF 000070304). After calcination at 800 °C (Fig. 5.1b), the basal spacing of Montmorillonite collapsed from 1.85 to 0.97 nm (2θ of 4.84°–9.13°) due to dehydration and dehydroxylation. The two new diffraction peaks that appeared at 2θ of 18.47° and 27.87° (Fig. 5.1b) correspond to the formation of $NaMgAlSi_4O_{11}$. The initial transformation process

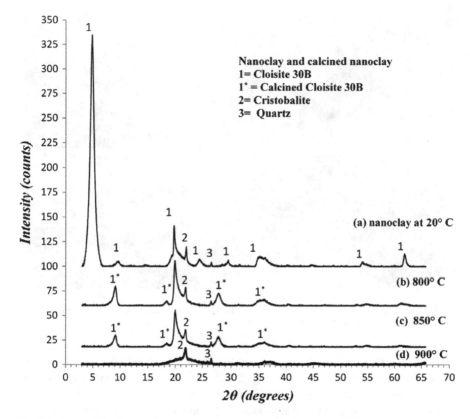

Fig. 5.1 X-ray diffraction patterns of nanoclay and calcined nanoclay [19]

of nanoclay was mainly due to the dehydration and dehydroxylation of montmorillonite clay. After further calcination at 850 °C (Fig. 5.1c) the basal spacing of Montmorillonite collapsed further to 0.96 nm. Finally, the peaks belonging to $NaMgAlSi_4O_{11}$ disappeared completely at 900 °C due to the destruction of its platelets and the concomitant formation of an amorphous phase of alumino-silicate (Fig. 5.1d).

5.2.2 High Resolution Transmission Electron Microscopy

HRTEM images of nanoclay (Cloisite 30B) are shown in Fig. 5.2a–b. The lower magnification image in Fig. 5.2a gives a general view of the nanoclay platelets. The high magnification image in Fig. 5.2b shows the layer structure of platelets. It can be seen clearly that the distances between the nanoclay platelets were about 1.85 nm and thus this is evidence that the d-spacing of (001) planes in nanoclay was 1.85 nm as shown in Table 2.3 in Chap. 2. However, Fig. 5.2c–d shows the

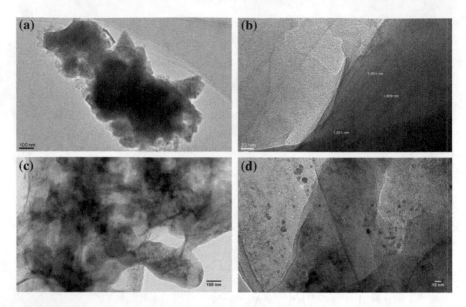

Fig. 5.2 TEM images of nanoclay and calcined nanoclay (at 900 °C) at low magnification (**a, c**), and high magnification (**b, d**) [19]

HRTEM images for calcined nanoclay (at 900 °C) at low and high magnification, respectively. At high magnification (Fig. 5.2d), it can be seen that many platelets in calcined nanoclay have been destroyed and broken into nanoparticles with semi-spherical shapes. The transformation of nanoclay platelets into amorphous alumino-silicate during calcination was due to: (a) dehydration and dehydroxylation of montmorillonite clay, (b) decomposition of quaternary ammonium salt, and (c) phase destruction of $NaMgAlSi_4O_{11}$. The natural montmorillonite clay [(Ca, Na)$_{0.3}$Al$_2$(Si,Al)$_4$O$_{10}$(OH)$_2$·xH$_2$O] has a 2:1 layer crystal structure that consists of aluminium octahedrons within two silicon tetrahedron layers. Dehydration can cause the loss of interlayer H_2O at low temperature and dehydroxylation can lead to OH removal from the octahedral sheets at higher temperatures. He et al. [20] indicated that when montmorillonite nanoclay was calcined at 920 °C, it transformed to an amorphous phase of alumina-silicate. Similarly, Shebl et al. [21] reported that the calcination of montmorillonite nanoclay (Cloisite 30B) at 850 °C for 2 h led to transformation from crystalline nano alumino-silicate into amorphous nano alumino-silicate.

5.2.3 Surface Morphology of Hemp Fabric

SEM micrographs of un-treated and NaOH-treated hemp fabrics are shown in Fig. 5.3a–b. It is clearly seen that the un-treated fabric had some impurities on its

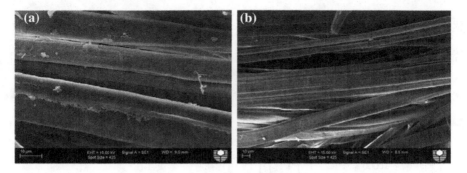

Fig. 5.3 SEM images showing the surface structure of: **a** untreated hemp fabric, **b** NaOH-treated hemp fabric [19]

surface (Fig. 5.3a), which were mostly waxes or fatty substances. However, after chemical treatment with NaOH solution (Fig. 5.3b), it can be seen that the fabric surface became more uniform due to the removal of these waxes or fatty substances [22, 23].

5.2.4 Crystallinity Index of Hemp Fabric

Figure 5.4 shows the XRD patterns of un-treated and NaOH-treated hemp fabrics. The X-ray diffraction patterns of hemp fabric show a typical crystal lattice of native cellulose (cellulose I). The fibre crystallinity index (*CrI*) of hemp fabric was determined by using the equation of the Segal empirical method [24, 25];

$$CrI = \frac{I_{002} - I_{am}}{I_{002}} \times 100 \tag{5.1}$$

where I_{002} is the maximum intensity of the (002) crystalline peak and I_{am} is the minimum intensity of the amorphous material between (101) and (002) peaks as shown in Fig. 5.4. The crystallinity index of un-treated and treated hemp fabric was found to be about 82.6 and 86.2%, respectively. It is clear that the NaOH treatment had increased the crystallinity index of hemp fabric as a result of the removal of hemicellulose, pectins, oils and waxes from the surface of hemp fabric [25]. Troedec et al. [23] reported that NaOH treatment is well-known to bleach and clean the surface of hemp fibres and to remove amorphous materials such as hemicellulose, pectins and impurities (fatty substances and waxes) from their surface. They observed that the cellulose crystallinity index of un-treated and NaOH treated hemp fibres was increased from 80 to 86%.

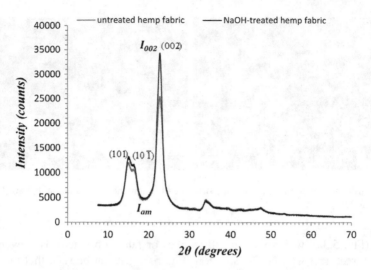

Fig. 5.4   X-ray diffraction patterns of untreated hemp fabric and NaOH-treated hemp fabric [19]

5.2.5   Quantitative X-Ray Diffraction Analysis of Nano-Matrix

The XRD patterns of cement paste and nanocomposites containing 1, 2 and 3 wt% CNC are shown in Fig. 5.5a–d, with Corundum [Al_2O_3] (PDF 000461212) as the internal standard. Table 5.1 shows the Rietveld quantitative phase analysis of cement paste and nanocomposites. Three important phases are noticed in this study: portlandite [$Ca(OH)_2$] (PDF 00-044-1481), tricalcium silicate [C_3S] (00-049-0442) and dicalcium silicate [C_2S] (PDF 00-033-0302). Moreover, four less important phases are also noticed: Ettringite [$Ca_6Al_2(SO_4)_3(OH)_{12}\cdot26H_2O$] (PDF 000411451), Gypsum [$Ca(SO_4)(H_2O)_2$] (PDF 040154421), Quartz [SiO_2] (PDF 000461045) and Calcite [$CaCO_3$] (PDF 000050586).

As can be seen from Table 5.1 and Fig. 5.5b, the addition of 1 wt% CNC reduced the amount of $Ca(OH)_2$ from 16.8 to 12.1 wt%, about 28% reduction when compared to cement paste. Also the intensities of major peaks of $Ca(OH)_2$ were significantly reduced when compared to cement paste (Fig. 5.5a, b). Furthermore, the amorphous content increased from 70.1 to 74.8 wt%, about 6.7% increase. This indicates an obvious consumption of $Ca(OH)_2$ crystals for the pozzolanic reaction due to the presence of CNC and good dispersion of calcined nanoclay in the matrix. As a result more amorphous calcium silicate hydrate gel (C-S-H) was produced. This explanation can be confirmed by the inspection of the amounts of unreacted C_3S (2.0 wt%) and C_2S (6.6 wt%), in which the amounts of unreacted C_3S and C_2S are slightly higher than the cement paste. Wei et al. [26] reported that pozzolanic reaction decelerates the hydration reaction of C_3S and C_2S during the curing time. In this study, these unreacted C_3S and C_2S could react with water later to produce

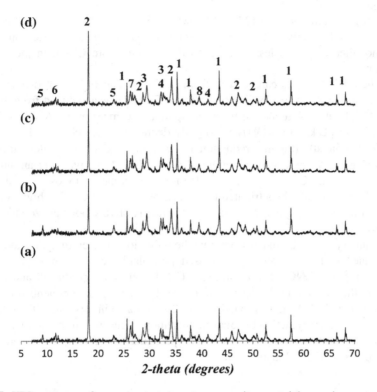

Fig. 5.5 XRD patterns of: **a** cement paste, nanocomposites containing various amounts of calcined nanoclay, **b** 1 wt% (CNCC1), **c** 2 wt% (CNCC2), **d** 3 wt% (CNCC3). *Legend* 1 = Corundum [Al_2O_3], 2 = Portlandite [$Ca(OH)_2$], 3 = Tricalcium silicate [C_3S], 4 = Dicalcium silicate [C_2S], 5 = Ettringite, 6 = Gypsum, 7 = Quartz, 8 = Calcite [19]

Table 5.1 QXDA results (Phase abundance) by Bruker *DIFFRAC^{plus}* TOPAS 4.2 software for cement paste and nanocomposites containing 1, 2 and 3 wt% calcined nanoclay (values in parentheses are the estimated standard deviation of the least significant figure) [19]

Weight % (Phase abundance)				
Phase	C	CNCC1	CNCC2	CNCC3
Portlandite [$Ca(OH)_2$]	16.8 (6)	12.1 (6)	13.2 (6)	14.1 (6)
Ettringite [$Ca_6Al_2(SO_4)_3(OH)_{12}\cdot26H_2O$]	2.0 (3)	1.3 (2)	1.5 (2)	1.8 (3)
Tricalcium silicate [C_3S]	1.3 (2)	2.0 (2)	1.7 (2)	1.4 (2)
Dicalcium silicate [C_2S]	4.4 (2)	6.6 (3)	6.1 (3)	5.4 (3)
Gypsum [$Ca(SO_4)(H_2O)_2$]	0.7 (1)	0.4 (1)	0.6 (1)	0.4 (1)
Calcite [$CaCO_3$]	3.7 (3)	2.1 (2)	2.7 (2)	3.3 (2)
Quartz [SiO_2]	0.9 (1)	0.6 (1)	0.4 (1)	0.7 (1)
Amorphous content	70.1 (8)	74.8 (7)	73.7 (7)	72.8 (8)
Rwp	5.28	5.22	5.29	5.17
Rexp	3.53	3.54	3.52	3.52
x^2 (Rwp/Rexp)	1.50	1.47	1.50	1.47

more C-S-H gel after 56 days [27, 28]. Recently, Shaikh et al. [29] reported that the quantitative XRD analysis after 28 days showed that the cement paste containing 2% nano-silica exhibited less calcium hydroxide but more C_2S than the control cement paste.

On the other hand, as can be seen from Table 5.1 and Fig. 5.5d for nanocomposites containing 3 wt% CNC, the amount of $Ca(OH)_2$ was reduced from 16.8 to 14.1 wt%, about 16% reduction when compared to cement paste. Also the intensities of major peaks of $Ca(OH)_2$ are slightly decreased when compared to cement paste (Fig. 5.5a, d). But this reduction of amount of $Ca(OH)_2$ is less than the reduction in nanocomposites containing 1 wt% CNC. Moreover, the amounts of C_3S (1.4 wt%) and C_2S (5.4 wt%) are also lower than nanocomposites containing 1 wt% CNC. This may be attributed to the agglomeration of CNC at high contents which led to relatively poor dispersion and hence relatively poor pozzolanic reaction [30].

In summary, it is important to note that the reduction of calcium hydroxide could be attributed to two reasons: (i) increased pozzolanic reaction from amorphous nanoparticles (i.e. CNC) that led to more C-S-H gel being produced, and (ii) reduction of the hydration reaction rate of the Portland cement components due to the pozzolanic reaction. As it was found in [26, 29] and also in this study, the effect of enhanced pozzolanic reaction by amorphous nanoparticles was more dominant than the reduction of the hydration reaction rate, particularly at the optimum content of nanoparticles with good dispersion.

5.2.6   Porosity and Density

The porosity, water absorption and density of cement paste, un-treated hemp fabric-reinforced cement composites (UHFRC), 6THFRC composites, nanocomposites and 6THFR-CNCC nanocomposites are shown in Table 5.2. As expected for all UHFRC composites the porosity and water absorption are increased and the density is decreased with increasing hemp fabric content as compared to cement paste [17]. However, it can be seen that the NaOH treatment of hemp fabric has slightly reduced the porosity and water absorption and increased the density of 6THFRC composite when compared to 6UHFRC. This slight improvement could be attributed to reduced voids in the fibre–matrix interface region after NaOH treatment in 6THFRC composite. Table 5.2 also shows that the addition of CNC decreases the porosity and water absorption of these composites when compared to control cement paste and 6THFRC composites. In 6THFR-CNCC1 composite, the porosity and water absorption decreased by 12.4 and 14%, respectively compared to 6THFRC composites. This indicates that CNC has filling effect in the porosity of cement paste composites with 6 treated hemp fabric [31]. Supit and Shaikh [32] reported that the addition of 2 wt% NS (nano-silica) significantly reduced the porosity of high volume fly ash (HVFA) concrete. Furthermore, In Table 5.2, the addition of 1 wt% CNC increased the density of control cement paste and 6THFRC

Table 5.2 Porosity and density values for cement paste (C), untreated (UHFRC) composites and 6THFRC composites, nanocomposites (CNCC) and treated hemp fabric-reinforced calcined nanoclay-cement nanocomposites (6THFR-CNCC) [19]

Sample	Porosity (%)	Density (10^3 kg/m^3)	Water absorption (%)
C	24.0 ± 0.5	1.76 ± 0.02	13.4 ± 0.7
4UHFRC	30.1 ± 0.7	1.61 ± 0.01	18.7 ± 0.6
5UHFRC	31.6 ± 0.7	1.56 ± 0.03	20.3 ± 0.7
6UHFRC	33.0 ± 0.4	1.53 ± 0.03	21.6 ± 0.6
7UHFRC	34.2 ± 0.6	1.51 ± 0.01	22.7 ± 0.5
6UHFRC	32.1 ± 0.8	1.55 ± 0.01	20.7 ± 0.7
CNCC1	16.5 ± 0.6	1.93 ± 0.01	8.9 ± 0.6
CNCC2	17.6 ± 0.5	1.91 ± 0.02	9.6 ± 0.8
CNCC3	18.9 ± 0.6	1.85 ± 0.02	10.3 ± 0.4
6THFR-CNCC1	28.1 ± 0.5	1.62 ± 0.01	17.8 ± 0.4
6THFR-CNCC2	29.4 ± 0.8	1.60 ± 0.01	19.0 ± 0.7
6THFR-CNCC3	30.2 ± 0.7	1.57 ± 0.03	19.8 ± 0.3

composite by 9.7 and 4.5%, respectively. This improvement demonstrated that cement composites with 1 wt% CNC yields more consolidated microstructure. However, the addition of more CNC leads to increase in porosity and water absorption and also decrease in density. This could be attributed to the poor dispersion and agglomerations of the CNC which create more voids in the matrix [31].

SEM examinations of the microstructure of control cement paste, nanocomposites containing 1 and 3 wt% CNC are shown in Fig. 5.6a–c. The SEM micrograph of control cement matrix (Fig. 5.6a) shows more Ca(OH)$_2$ crystals (portlandite), Ettringite and pores when compared to nanocomposites. Figure 5.6b shows the SEM micrograph of nanocomposites containing 1 wt% CNC, where the microstructure appears denser with fewer pores and more C-S-H gels than the control cement matrix. On the other hand, the nanocomposite containing 3 wt% CNC (Fig. 5.6c) shows more pores than 1 wt% CNC nanocomposite. These SEM results confirm the reduction of portlandite crystals in nanocomposites when compared to cement matrix, thus in agreement with the quantitative X-ray diffraction results (QXDA) above.

5.2.7 Flexural Strength of Hemp Fabric-Reinforced Cement Composites

Values of flexural strength for cement paste, un-treated hemp fabric-reinforced cement composite (UHFRC) and 6THFRC composites are shown in Fig. 5.7. It can be seen that the flexural strengths of all hemp fabric-reinforced cement composites have significantly improved when compared to cement paste. This enhancement in flexural properties can be attributed to the ability of hemp fabric to withstand the

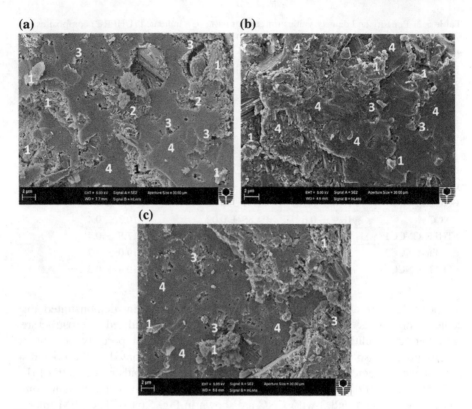

Fig. 5.6 SEM micrographs of: **a** cement paste, **b** nanocomposites containing 1 wt% calcined nanoclay, **c** nanocomposites containing 3 wt% calcined nanoclay. *Legend* 1 = [Ca(OH)$_2$] crystals, 2 = Ettringite, 3 = pores, 4 = C-S-H gel [19]

bending force. Peled and Bentur [13] studied the effect of High Density Polyethylene (HDP) with 8 fabric layers on the flexural strength of cement composites. They reported that the flexural strength of HDP fabric-reinforced cement composite was increased by about 173.7% when compared to control cement paste.

In this study, the flexural strength was increased with an increase in fibre content up to the optimum hemp fibre content, and then decreased after this limit. The optimum hemp fabric content was found to be 6.9 wt%, in which the flexural strength increased from 5.4 to 12.6 MPa, about 133% increase when compared to cement paste. However, beyond this optimum content of hemp fibre, the flexural strength of the UHFRC composites decreased due to the poor adhesion between the fibres and the matrix [33]. For example, the flexural strength of 7UHFRC composites with 8.1 wt% hemp fabric content decreased by about 11.5% when compared to 6UHFRC composites. In a similar work, Bentchikou et al. [34] studied recycled cellulose fibres-cement board with fibre fraction ranged from 0 to 16 wt%. They concluded that composite with the optimum fibre content (4 wt%) gave the

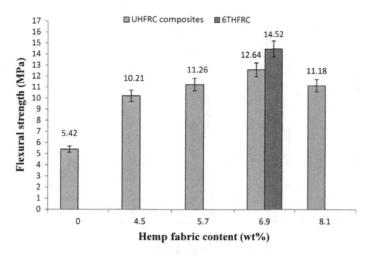

Fig. 5.7 Flexural strength as a function of hemp fabric content for cement paste, untreated hemp fabric-reinforced cement composites (UHFRC) and treated hemp fabric-reinforced composites (6THFRC) [19]

maximum flexural strength. In addition, Fig. 5.7 also shows the effect of NaOH treatment of hemp fabric on the flexural strength of 6THFRC composites. It can be clearly seen that the flexural strength of 6THFRC composites has increased from 12.6 to 14.5 MPa, about 14.9% increase when compared to 6UHFRC composite. This improvement may be explained as follows: after NaOH treatment, most of waxes and fats were removed from the hemp fibre surface, and the surface became more uniform but rough. Thus, this could lead to good interfacial bonding between the matrix and the hemp fibres which serves to enhance the load transfer process at the interface [35]. Sedan et al. [3] studied the untreated and treated hemp fibre reinforced cement composites with different fibre volume fractions of 7, 10, 16 and 20 vol.% (w/c = 0.5). They reported that the flexural strength of NaOH treated hemp fibre reinforced cement composite with the optimum fibre content of 16 vol.% reached up to 9.5 MPa.

The flexural strengths of nanocomposites and treated hemp fabric reinforced-nanocomposites are shown in Fig. 5.8. Overall, the incorporation of CNC into the cement matrix has led to significant enhancement in the flexural strengths of all nanocomposites and treated hemp fabric reinforced nanocomposites. The flexural strength of nanocomposites containing 1, 2 and 3 wt% CNC increased by 42.9, 34.8 and 30.6% respectively when compared to cement paste. In addition, the flexural strength of treated hemp fabric reinforced nanocomposites containing 1 wt% CNC (6THFR-CNCC1) increased from 14.5 to 20.2 MPa, about 38.8% increase when compared to treated hemp fabric-reinforced composites. This improvement clearly indicates the effectiveness of CNC in consuming calcium hydroxide (CH), supporting pozzolanic reaction, and filling the micro pores in the matrix. Thus the microstructure of nanocomposite matrix was denser than the cement matrix [21, 36].

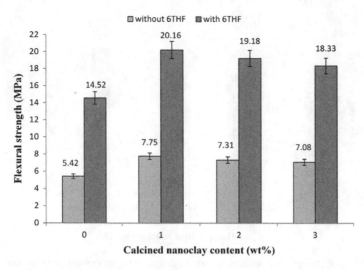

Fig. 5.8 Flexural strength as a function of calcined nanoclay content for cement paste and its nanocomposites with and without treated hemp fabric (6THF) [19]

Consequently, the interfacial bonding of treated hemp fabric-nanocomposite matrix was mostly improved, especially in the case of using 1 wt% CNC, as evident from its higher flexural strength. An analogous research was done by Khorami and Ganjian [37] where they studied the bagasse fibre-reinforced cement matrix with fibre content of 4 wt%, and silica fume was used as 5% replacement for cement. They observed that the flexural strength increased by about 20% when compared to control bagasse fibre-reinforced cement matrix. They attributed this improvement to the Pozzolanic and filler effects of very fine silica fume particles, which led to enhancement of the bonding strength between the matrix and fibres. However, the addition of CNC with more than 1 wt% caused a marked decrease in flexural strength. This strength reduction could be attributed to relatively poor dispersion and agglomeration of the CNC in the cement matrix at higher CNC contents, which caused an increase in porosity and led to the poor adhesion between the fibres and the matrix [38, 39]. Nevertheless the addition of CNC improved the flexural strength of treated hemp fabric reinforced cement composites. For example, in this study, although the flexural strength of composite with 3 wt% CNC was decreased when compared to composite with 1 wt% CNC but it was still higher than the control composite.

The stress-midspan deflection curves for treated hemp fabric reinforced cement composite and treated hemp fabric reinforced nanocomposites containing 1, 2 and 3 wt% CNC are shown in Fig. 5.9. The treated hemp fabric reinforced nanocomposite containing 1 wt% CNC shows the highest stress. This is due to high fibre-matrix interfacial bonding, which increases the maximum load-transfer capacity. On the other hand, the treated hemp fabric reinforced nanocomposites containing 2 and 3 wt% CNC and treated hemp fabric reinforced cement composite

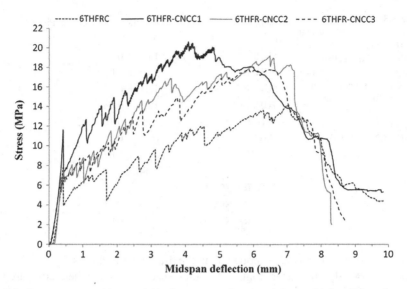

Fig. 5.9 Stress versus mid-span deflection curves for treated hemp fabric-reinforced cement composites and treated hemp fabric-reinforced nanocomposites [19]

show low flexural stress. This could be attributed to the increase in porosity which reduced the bond strength between the fibres and the matrix, and thus the load-transfer capacity.

5.2.8 Fracture Toughness

Results of fracture toughness for cement paste, untreated hemp fabric-reinforced cement composites (UHFRC) and treated hemp fabric-reinforced composites (6THFRC) are shown in Table 5.3. Overall, these composites showed significant improvement in fracture toughness. This enhancement can be attributed to fracture resistance provided by the hemp fabric which resulted in increased energy dissipation from crack-deflection at the fibre–matrix interface, fibre-debonding, fibre-bridging, fibre pull-out and fibre-fracture [40, 41]. As such, these composites are likely to exhibit crack-growth resistance or R-curve behaviour in their fracture resistance due to substantial fibre-bridging at the crack-wake. In UHFRC composites, the 6UHFRC composite achieved the highest fracture toughness with improvement reaching up to 303% when compared to cement paste. However, the increase of the fibre content beyond the optimum content led to a decrease in fracture toughness, as indicated by the lower fracture toughness for the 7UHFRC composites. After NaOH treatment, the 6THFRC composite exhibited 13.5% increase in fracture toughness. This result confirms that chemical treatment has improved the interfacial bond between the matrix and the treated hemp fibres. In a

Table 5.3 Fracture toughness values for cement paste (C), untreated (UHFRC) composites and 6THFRC composites, nanocomposites (CNCC) and treated hemp fabric-reinforced calcined nanoclay-cement nanocomposites (6THFR-CNCC) [19]

Sample	Fracture toughness (MPa m$^{1/2}$)
C	0.35 ± 0.02
4UHFRC	1.07 ± 0.11
5UHFRC	1.26 ± 0.09
6UHFRC	1.41 ± 0.11
7UHFRC	1.23 ± 0.08
6THFRC	1.60 ± 0.10
CNCC1	0.49 ± 0.02
CNCC2	0.47 ± 0.03
CNCC3	0.44 ± 0.03
6THFR-CNCC1	2.21 ± 0.10
6THFR-CNCC2	2.14 ± 0.11
6THFR-CNCC3	2.04 ± 0.09

similar study, Li et al. [42] reported that the fracture toughness of treated sisal textile reinforced vinyl-ester composites increased by 31% when compared to untreated ones.

Values of fracture toughness for nanocomposites with and without treated hemp fabrics are also shown in Table 5.3. The addition of CNC into treated hemp fabric reinforced nanocomposites significantly increased the fracture toughness. The fracture toughness of treated hemp fabric reinforced nanocomposites containing 1, 2 and 3 wt% CNC was 2.21, 2.14 and 2.04 MPa m$^{1/2}$, respectively. It can again be seen that the fracture toughness of 6THFR-CNCC1 composite was increased by 38.1% which can be attributed to the fact that CNC modified the matrix through pozzolanic reaction and reduction of $Ca(OH)_2$ content. Thus, good interfacial bonding between the nanomatrix and treated hemp fibres was achieved. In a similar study, Alamri and Low [43] reported that the addition of 1 wt% halloysite nanotubes (HNTs) into recycled cellulose fibres (RCF)/epoxy matrix significantly increased the fracture toughness by 38.8%. However, when the CNC content increased over the optimum content of 1 wt%, the facture toughness of nanocomposites and treated-hemp fabric-reinforced nanocomposites gradually decreased. This can be attributed to the poor dispersion of CNC within the matrix, which leads to an increase in porosity and weakening of the interfacial bond between the fibres and the matrix.

5.2.9   Failure Mechanisms

Figure 5.10a–f shows the SEM micrographs of 6UHFRC composite, 6THFRC composites and treated hemp fabric-reinforced nanocomposites containing 1 and 3 wt% CNC. A variety of failure mechanisms such as fibre-matrix interfacial debonding, fibre pull-out, rupture fibre and matrix fracture are observed. SEM

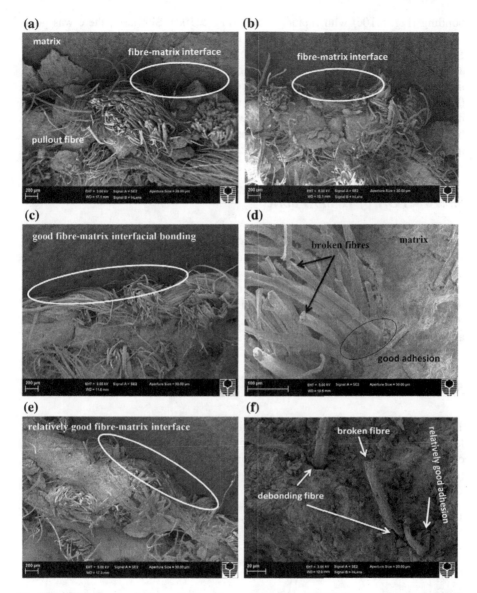

Fig. 5.10 SEM images showing the fracture surfaces of: **a** 6UHFRC composite, **b** 6THFRC composite, **c, d** 6THFR-CNCC1 nanocomposite, and **e, f** 6THFR-CNCC3 nanocomposite [19]

micrographs of 6UHFRC composite (Fig. 5.10a) shows relatively poor fibre-matrix interfaces with small gaps between the fabric layers and the cement matrix. In contrast, the 6THFRC composite (Fig. 5.10b) shows very small gaps between the fabrics and the matrix, which indicates better fibre-matrix interfacial bonding. The images of 6THFR-CNCC1 nanocomposite show good fibre-nanomatrix interfacial bonding (Fig. 5.10c) with ruptured fibres (Fig. 5.10d). Similarly, there was good adhesion between the fibre and the matrix in the 6THFR-CNCC3 nanocomposite as evidenced by broken fibres and debonding of fibres (Fig. 5.10e, f). Relatively good adhesion between the fibre and the matrix and fibre ruptures indicates that the fibre-matrix interface in this nanocomposite was better than the 6THFRC composite (Fig. 5.10b, e). In addition, these failure mechanisms of composites are in agreement with the stress versus mid-span deflection curves (Fig. 5.9). For example, 6THFR-CNCC1 nanocomposite shows the highest flexural stress (Fig. 5.9) which indicates better fibre-nanomatrix interfacial bonding as shown in (Fig. 5.10c).

5.2.10 TGA of Hemp Fabric

The thermograms (TGA) of untreated hemp fabric and treated hemp fabric are shown in Fig. 5.11. It can be seen from TGA curve that the weight loss (%) between 285 and 390 °C is due to decomposition of cellulose [44]. Among all range 25–1000 °C, it can be seen that the treated hemp fabric shows slightly higher thermal stability than untreated. This indicates that the NaOH treatment increases the thermal stability of hemp fibres through removal of the most fats, waxes and amorphous materials [9, 23].

5.2.11 Impact Strength

The impact strength is defined as the ability of the material to withstand impact loading [45]. Figure 5.12 shows the impact strength of cement paste, 6UHFRC composites, 6THFRC composites and nanocomposites with treated hemp fabrics (6THF). Generally, it can be seen that the impact strength of cement paste is

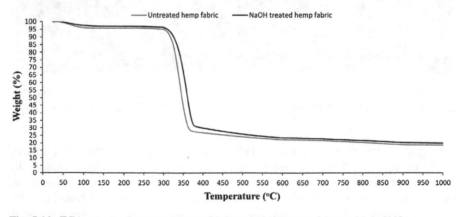

Fig. 5.11 TGA curves of untreated hemp fabric and NaOH-treated hemp fabric [19]

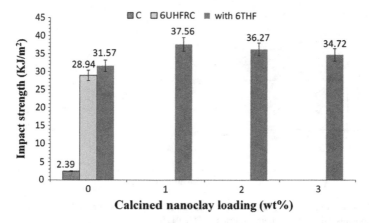

Fig. 5.12 Impact strength as a function of calcined nanoclay content for cement paste and 6UHFRC composite, 6THFRC composite and nanocomposites with treated hemp fabrics (6THF) [19]

significantly improved due to reinforced by Hemp fabrics [41]. It can be seen clearly that after NaOH treatment, the impact strength of 6THFRC composite has slightly increased from 28.94 to 31.57 by about 9% compared to 6UHFRC composite. This result indicates that relatively good interfacial bonding between the Hemp fabric and the cement matrix was achieved due to NaOH treatment of Hemp fabric. As shown in Fig. 5.12, the presence of CNC enhanced the impact strength for treated hemp fabric-reinforced nanocomposites.

The impact strength of 6THFR-CNCC1 was 37.56 kJ/m^2, about 19% increase compared to 6HFRC composite. This is due to good interfacial bonding between the fibres and the nanomatrix. Alhuthali et al. [46] reported that the addition of 3 wt% nanoclay into recycled cellulose fibres (RCF)/vinyl ester matrix increased the impact strength by 27% compared to RCF-reinforced vinyl ester composites. However, as CNC loading increased after the optimum content of 1 wt% the impact strength is decreased. For example, the impact strength of 6THFR-CNCC3 was 34.72 kJ/m^2, about 10% increase compared to 6HFRC composite, in which it is less than 6THFR-CNCC1. This reduction in impact strength at higher CNC loading was due to the formation of CNC agglomerates and voids which led to reduced fibre–nanomatrix adhesion [47].

5.2.12   Thermal Stability

The thermograms (TGA) of cement paste, 6UHFRC composites, 6THFRC composites and treated hemp fabric reinforced-nanocomposites (6THFR-CNCC) are shown in Fig. 5.13. The char yields at different temperatures are summarized in Table 5.4. The TGA analysis shows four distinct stages of decomposition in these samples. The first stage of decomposition is between room temperature and 230 °C,

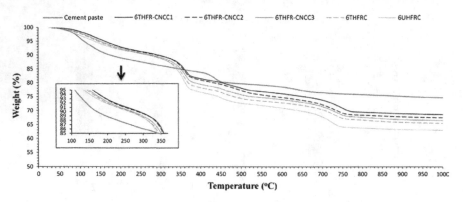

Fig. 5.13 TGA curves of cement paste and 6UHFRC composite, 6THFRC composite and nanocomposites with treated hemp fabrics (6THF) [19]

Table 5.4 Thermal properties of cement paste (C), 6UHFRC, 6THFRC and 6THFR-CNCC nanocomposites [19]

Sample	Char yield (%) at different temperature (°C)									
	100	200	300	400	500	600	700	800	900	1000
C	95.82	88.82	86.03	83.57	79.57	78.44	76.37	75.69	75.09	74.61
6UHFRC	96.94	91.81	88.51	76.46	72.56	70.88	67.58	63.80	63.25	62.94
6THFRC	97.56	91.91	88.57	78.13	74.23	72.75	70.21	66.58	65.81	65.51
6THFR-CNCC1	98.38	92.76	89.53	81.32	77.93	75.99	73.66	69.40	68.86	68.57
6THFR-CNCC2	98.31	92.68	89.43	80.85	77.08	74.63	71.99	68.34	67.77	67.48
6THFR-CNCC3	97.81	92.20	88.85	79.32	75.58	73.90	71.34	67.58	66.73	66.49

which may be related to the decomposition of Ettringite and dehydration of C–S–H gel (loss of water). The second stage of decomposition is between 285 and 390 °C, which corresponds to decomposition of cellulose of hemp fibre. The third stage of decomposition is between 400 and 510 °C, which corresponds to $Ca(OH)_2$ decomposition. The last stage of decomposition is between 670 and 780 °C, which correspond to $CaCO_3$ decomposition [48–50].

In the first stage, generally all composites with Hemp fabrics showed better thermal stability than cement paste due to resistance of Hemp fabrics to the decomposition. Furthermore, 6THFR-CNCC composite exhibited slightly better thermal stability than 6UHFRC composite, 6THFRC composites and cement paste due to resistance of calcined nanoclay to the decomposition [46, 47]. The 6THFRC composite in the second, third and fourth stage showed higher thermal stability than 6UHFRC because of the efficiency of NaOH treatment. From Table 5.4 at 1000 °C, the char residue of 6UHFRC and 6THFRC composites was about 65.51 and 62.94 wt%, respectively. Hence, it can be said that the 6THFRC composite performed better in thermal stability with slightly higher char residue of about 4% more than 6UHFRC [51]. Concerning 6THFR-CNCC composites in second, third and fourth stage, the 6THFR-CNCC1 composites show better thermal stability than 6THFRC composites, 6THFR-CNCC2 composites and 6THFRCNCC3 composites due to dense and compact nanomatrix through consumption of calcium hydroxide and formation of secondary CSH gels during pozzolanic reaction [30]. Whereas, 6THFR-CNCC3 composites show lower thermal stability than 6THFR-CNCC1 composites but higher than 6THFRC composites, in which this result confirms that slightly poor pozzolanic reaction has occurred and hence nanomatrix is less compacted. From Table 5.4 at 1000 °C, the char residue of 6THFRC composites, 6THFR-CNCC1 composites, 6THFR-CNCC2 composites and 6THFR-CNCC3 composites was about 65.51, 68.57, 67.48 and 66.49 wt% respectively. It can be seen that 6THFR-CNCC1 composites performed better in thermal stability with higher char residue of about 5 and 3% more than 6UHFRC and 6THFR-CNCC3 composites, respectively. In a similar study, Chen et al. [52] reported that addition of 10 wt% nano-TiO_2 into cement paste improved the thermal stability of nanocomposite considerably. However, overall in between 360 and 1000 °C, the 6THFR-CNCC1 composite showed lower thermal stability than cement paste but still better than other samples.

5.3 Conclusions

The influence of calcined nanoclay (CNC) and chemical treatment of Hemp fabric on the microstructure and mechanical properties of treated hemp fabric-reinforced cement nanocomposites has been investigated. The optimum hemp fabric content for these nanocomposites is 6.9 wt% (i.e. 6 fabric layers). Alkali-treated hemp fabric-reinforced cement composites exhibit the highest flexural strength when compared to their non-treated counterparts. In addition, mechanical properties are improved as a result of CNC addition. An optimum replacement of ordinary Portland cement with 1 wt% CNC is observed through reduced porosity and increased density, flexural strength and fracture toughness in treated hemp fabric-reinforced nanocomposite. It is shown that CNC behaves not only as a filler to improve the microstructure, but also as the activator to facilitate the pozzolanic reaction and thus improved the adhesion between the treated hemp fabric and the matrix. The mechanical and thermal properties of NaOH treated hemp fabric reinforced cement composites were noticeably improved when compared to the nontreated counterparts. In the 6THFR-CNCC1 nanocomposites, the porosity and water absorption declined by 12.4 and 14% respectively, as well as the density, flexural strength, fracture toughness, impact strength and thermal stability improved by 4.5, 38.8, 38.1, 19 and 5% respectively compared to the 6THFRC composites. Indeed, particles agglomeration increased as CNC content increased which adversely reduced the mechanical properties of composites. It could be recommended that much research is required to overcome the CNC agglomerations. Construction applications of this eco-nanocomposite involve ceilings and roofing.

References

1. Silva F, Mobasher B. Cracking mechanisms in durable sisal fiber reinforced cement composites. Cem Concr Compos. 2009;31:721–30.
2. Islam S, Hussain R, Morshed M. Fiber-reinforced concrete incorporating locally available natural fibers in normal-and high-strength concrete and a performance analysis with steel fiber-reinforced composite concrete. J Compos Mater. 2012;46:111–22.
3. Sedan D, Pagnoux C, Smith A, Chotard T. Mechanical properties of hemp fibre reinforced cement: influence of the fibre/matrix interaction. J Eur Ceram Soc. 2008;28:183–92.
4. Ali M, Liu A, Sou H, Chouw N. Mechanical and dynamic properties of coconut fibre reinforced concrete. Constr Build Mater. 2012;30:814–25.
5. Elsaid A, Dawood M, Seracino R, Bobko C. Mechanical properties of kenaf fiber reinforced concrete. Constr Build Mater. 2011;25:1991–2001.
6. Awwada E, Mabsout M, Hamad B, Farran M, Khatib H. Studies on fiber-reinforced concrete using industrial hemp fibers. Constr Build Mater. 2012;35:710–7.
7. Li Z, Wang L, Wang X. Compressive and flexural properties of hemp fiber reinforced concrete. Fiber Polym. 2004;5:187–97.
8. Li Z, Wang X, Wang L. Properties of hemp fibre reinforced concrete composites. Compos A. 2006;37:497–505.
9. Rachini A, Troedec M, Peyratout C, Smith A. Comparison of the thermal degradation of natural, alkali-treated and silane-treated hemp fibers under air and an inert atmosphere. J Appl Polym Sci. 2009;112:226–34.
10. Blankenhorn P, Blankenhorn B, Silsbee M, DiCola M. Effects of fiber surface treatments on mechanical properties of wood fiber–cement composites. Cem Concr Res. 2001;31:1049–55.
11. Dalmay P, Smith A, Chotard T, Sahay-Turner P, Gloaguen V, Krausz P. Properties of cellulosic fibre reinforced plaster: influence of hemp or flax fibres on the properties of set gypsum. J Mater Sci. 2010;45:793–803.
12. Troèdec M, Peyratout C, Smith A, Chotard T. Influence of various chemical treatments on the interactions between hemp fibres and a lime matrix. J Eur Ceram Soc. 2009;29:1861–8.
13. Peled A, Bentur A. Fabric structure and its reinforcing efficiency in textile reinforced cement composites. Compos A. 2003;34:107–18.
14. Peled A, Sueki S, Mobasher B. Bonding in fabric–cement systems: effects of fabrication methods. Cem Concr Res. 2006;36:1661–71.
15. Peled A, Mobasher B. Tensile behavior of fabric cement-based composites: pultruded and cast. J Mater Civ Eng. 2007;19:340–8.
16. Soranakom C, Mobasher B. Geometrical and mechanical aspects of fabric bonding and pull out in cement composites. Mater Struct. 2009;42:765–77.
17. Alamri H, Low IM, Alothman Z. Mechanical, thermal and microstructural characteristics of cellulose fibre reinforced epoxy/organoclay nanocomposites. Compos B. 2012;43:2762–71.
18. Snoeck D, De Belie N. Mechanical and self-healing properties of cementitious composites reinforced with flax and cottonised flax, and compared with polyvinyl alcohol fibres. Biosyst Eng. 2012;111:325–35.
19. Hakamy A. Microstructural design of high-performance natural fibre nanoclay cement nanocomposites. Ph.D. Thesis, Curtin University, Perth Australia; 2016.
20. He C, Makovicky E, Osbaeck B. Thermal treatment and pozzolanic activity of Na- and Ca-montmorillonite. Appl Clay Sci. 1996;10:351–68.
21. Shebl S, Allie L, Morsy M, Aglan H. Mechanical behavior of activated nano silicate filled cement binders. J Mater Sci. 2009;44:1600–6.
22. Troèdec M, Rachini A, Peyratout C, Rossignol S, Max E, Kaftan O, Fery A, Smith A. Influence of chemical treatments on adhesion properties of hemp fibres. J Colloid Interface Sci. 2011;356:303–10.

23. Troedec M, Sedan D, Peyratout C, Bonnet J, Smith A, Guinebretiere R, Gloaguen V, Krausz P. Influence of various chemical treatments on the composition and structure of hemp fibres. Compos A. 2008;39:514–22.

24. Sedan D, Pagnoux C, Chotard T, Smith A, Lejolly D, Gloaguen V. Effect of calcium rich and alkaline solutions on the chemical behaviour of hemp fibres. J Mater Sci. 2007;42:9336–42.

25. Sawpan M, Pickering K, Fernyhough A. Effect of various chemical treatments on the fibre structure and tensile properties of industrial hemp fibres. Compos A. 2011;42:888–95.

26. Wei Y, Yao W, Xing X, Wu M. Quantitative evaluation of hydrated cement modified by silica fume using QXRD, 27Al MAS NMR, TG–DSC and selective dissolution techniques. Constr Build Mater. 2012;36:925–32.

27. Scrivenera K, Fullmanna T, Galluccia E, Walentab G, Bermejob E. Quantitative study of Portland cement hydration by X-ray diffraction/Rietveld analysis and independent methods. Cem Concr Res. 2004;34:1541–7.

28. Govindarajan D, Gopalakrishnan R. Spectroscopic studies on Indian Portland cement hydrated with distilled water and sea water. Front Sci. 2011;1:21–7.

29. Shaikh FUA, Supit SWM, Sarker PK. A study on the effect of nano silica on compressive strength of high volume fly ash mortars and concretes. Mater Des. 2014;60:433–42.

30. Soin A, Catalan L, Kinrade S. A combined QXRD/TG method to quantify the phase composition of hydrated Portland cements. Cem Concr Res. 2013;48:17–24.

31. Jo B, Kim C, Tae G, Park J. Characteristics of cement mortar with nano-SiO₂ particles. Constr Build Mater. 2007;21:1351–5.

32. Supit SWM, Shaikh FUA. Durability properties of high volume fly ash concrete containing nano-silica. Mater Struct. 2014;. doi:10.1617/s115270140329-0.

33. Abdullah A, Jamaludin S, Noor M, Hussin K. Composite cement reinforced coconut fiber: physical and mechanical properties and fracture behavior. Aus J Basic Appl Sci. 2011;5:1228–40.

34. Bentchikou M, Guidoum A, Scrivener K, Silhadi K, Hanini S. Effect of recycled cellulose fibres on the properties of lightweight cement composite matrix. Constr Build Mater. 2012;34:451–6.

35. Asprone D, Durante M, Prota A, Manfredi G. Potential of structural pozzolanic matrix–hemp fiber grid composites. Constr Build Mater. 2011;25:2867–74.

36. Chang T, Shih J, Yang K, Hsiao T. Material properties of portland cement paste with nano-montmorillonite. J Mater Sci. 2007;42:7478–87.

37. Khorami M, Ganjian E. Comparing flexural behaviour of fibre-cement composites reinforced bagasse: wheat and eucalyptus. Constr Build Mater. 2011;25:3661–7.

38. Filippi S, Paci M, Polacco G, Dintcheva N, Magagnini P. On the interlayer spacing collapse of Cloisite® 30B organoclay. Polym Degrad Stab. 2011;96:823–32.

39. Li H, Xiao H, Yuan J, Ou J. Microstructure of cement mortar with nano-particles. Compos B. 2004;35:185–9.

40. Ahmed SFU, Mihashi H. Strain hardening behavior of lightweight hybrid polyvinyl alcohol (PVA) fiber reinforced cement composites. Mater Struct. 2011;44:1179–91.

41. Zhou X, Ghaffar S, Dong W, Oladiran O, Fan M. Fracture and impact properties of short discrete jute-fiber reinforced cementitious composites. Mater Des. 2013;49:35–47.

42. Li Y, Mai Y, Ye L. Effects of fibre surface treatment on fracture-mechanical properties of sisal-fibre composites. Compos Interface. 2005;12:141–63.

43. Alamri H, Low IM. Microstructural, mechanical, and thermal characteristics of recycled cellulose fiber-Halloysite-epoxy hybrid composites. Polym Compos. 2012;33:589–600.

44. Le Troëdec M, Peyratout C, Smith A, Chotard T. Influence of various chemical treatments on the interactions between hemp fibres and a lime matrix. J Eur Ceram Soc 2009;29:1861–68.

45. Toutanji H, Xu B, Gilbert J, Lavin T. Properties of poly (vinyl alcohol) fiber reinforced high-performance organic aggregate cementitious material: converting brittle to plastic. Constr Build Mater. 2010;24:1–10.

46. Alhuthali A, Low IM, Dong C. Characterization of the water absorption, mechanical and thermal properties of recycled cellulose fibre reinforced vinylester eco-nanocomposites. Compos B. 2012;43:2772–81.
47. Hakamy A, Shaikh FUA, Low IM. Thermal and mechanical properties of hemp fabric-reinforced nanoclay–cement nanocomposites. J Mater Sci. 2014;49:1684–94.
48. Filho J, Silva F, Filho R. Degradation kinetics and aging mechanisms on sisal fiber cement composite systems. Cem Concr Compos. 2013;40:30–9.
49. Lothenbach B, Winnefeld F, Alder C, Wieland E, Lunk P. Effect of temperature on the pore solution, microstructure and hydration products of Portland cement pastes. Cem Concr Res. 2007;37:483–91.
50. Djaknoun S, Ouedraogo E, Benyahia A. Characterisation of the behaviour of high performance mortar subjected to high temperatures. Constr Build Mater. 2012;28:176–86.
51. Beckermann G, Pickering K. Engineering and evaluation of hemp fibre reinforced polypropylene composites: fibre treatment and matrix modification. Compos A. 2008;39:979–88.
52. Chen J, Kou S, Poon C. Hydration and properties of nano-TiO_2 blended cement composites. Cem Concr Compos. 2012;34:642–9.

Chapter 6
Durability of Naoh-Treated Hemp Fabric and Calcined Nanoclay-Reinforced Cement Nanocomposites

6.1 Introduction

This chapter presents the durability of hemp fibre reinforced cement composites by combining both methods e.g. modification of fibre surfaces using NaOH solution and the cement matrix by adding calcined nano clay. The effect of calcined nanoclay (CNC) on the durability properties of treated hemp fabric-reinforced cement composite is studied. The durability of the treated hemp fabric-reinforced cement composites and nanocomposites is discussed based on the porosity and flexural strength obtained at 56 days and 236 days. Thus eight series of mixes are considered in this study. The first series is control matrix termed as "C" and consisted of 100% OPC. The second, third and fourth series are similar to the first series in every aspect except the partial replacement of OPC by 1, 2 and 3 wt% of CNC, respectively. These series are termed as CNCC1, CNCC2 and CNCC3, respectively. In fifth series the control OPC matrix is reinforced by NaOH treated 6 layers of HF and is termed as 6THFRC. The sixth, seventh and eighth series are similar to fifth series in every aspect except the partial replacement of OPC by 1, 2 and 3 wt% of CNC, respectively. These series are termed as 6THFRC-CNCC1, 6THFRC-CNCC2 and 6THFRC-CNCC3, respectively. A constant water/binder ratio of 0.485 is considered in all mixes.

6.2 Results and Discussion

6.2.1 Nanocomposites

The porosity of cement paste and nanocomposites are shown in Table 6.1. Overall, it can be seen that the porosity of C, CNCC1, CNCC2 and CNCC3 decreased slightly in periods of between 56 and 236 days. Firstly, it is known that after

© The Author(s) 2017
I.-M. Low et al., *High Performance Natural Fiber-Nanoclay Reinforced Cement Nanocomposites*, Biobased Polymers, DOI 10.1007/978-3-319-56588-0_6

90 days curing, the compressive strength of control concrete or cement paste increase slightly with increasing ages as a result of reduction in porosity due to further hydration reaction. Secondly, in nanocomposites, slight reduction in porosity with increasing ages can be attributed to both hydration reaction and pozzolanic reaction as well as filling effect. For example, in CNCC1 cement nanocomposite, the porosity at 56 days is decreased by 31.2% compared to cement paste, however at 236 days it decreased slightly from 16.5 to 15.1% by about 8% decrease. This indicated that nanocomposites containing 1, 2 and 3 wt% CNC had both filling and pozzolanic effects in the porosity of cement paste composites as well as to further hydration reaction (in between 56 and 236 days), in which the nanocomposite matrix becomes more a consolidated microstructure due to filling of the micro pores and densification by the enhanced pozzolanic activity and further hydration reaction [1]. This result is in agreement with the work done by Supit and Shaikh [2] reported that the addition of 2 wt% nano-silica significantly reduced the porosity of high volume fly ash (HVFA) concrete after 90 days. However, the addition of more CNC leads to increase in porosity. This could be attributed to the poor dispersion and agglomerations of the CNC which create more voids in the matrix [3].

The XRD patterns of cement paste and nanocomposites containing 1, 2 and 3 wt% CNC after 236 days are shown in Fig. 6.1. The addition of 1, 2, 3 wt% CNC reduced the intensities of major peaks of $Ca(OH)_2$ when compared to cement paste. For example intensity of $Ca(OH)_2$ at 2θ of $18.01°$ for CNCC1 is decreased by 13.9% compared to cement paste. This indicated an obvious consumption of $Ca(OH)_2$ in the pozzolanic reaction due to the presence of CNC and good dispersion of calcined nanoclay in the matrix which produce more amorphous calcium silicate hydrate gel (C-S-H). The formation of more C-S-H gel in CNCC1 nanocomposites can be also confirmed by the inspection of the increase in intensity peaks corresponding to 2θ of $31°$ (as mentioned by Hosino et al. [5]) in the close-up of Fig. 6.1 where peak for C_3S at 2θ of $29.4°$ is also reduced. Moreover, it can also be seen in the same close-up figure that the $Ca(OH)_2$ peak at 2θ of $28.7°$ is also lower in CNCC1 nanocomposite

Table 6.1 Porosity (%) values for cement paste (C), (CNCC) nanocomposites, 6THFRC composites and (6THFR-CNCC) nanocomposites at 56 and 236 days [4]

Sample	Porosity (%) at 56 days	Porosity (%) at 236 days
C	23.9 ± 0.5	22.3 ± 0.7
CNCC1	16.5 ± 0.6	15.1 ± 0.3
CNCC2	17.6 ± 0.5	16.2 ± 0.5
CNCC3	18.9 ± 0.6	17.3 ± 0.7
6THFRC	32.1 ± 0.8	36.3 ± 0.7
6THFR-CNCC1	28.1 ± 0.5	30.5 ± 0.9
6THFR-CNCC2	29.4 ± 0.8	31.8 ± 0.9
6THFR-CNCC3	30.2 ± 0.7	33.1 ± 1.0

than the cement paste due to pozzolanic reaction. In the same close-up figure the XRD for CNCC1 at 56 days (as reported by our previous work, Hakamy [4]) is also plotted. By comparing the peaks for CNCC1 at 56 and 236 days it can also be seen that the XRD peak of CNCC1 at 236 days at 2θ of $31°$ is slightly higher than at 56 days indicating the formation of more C-S-H and consumption of CH after 236 days. The reduction of C_3S peak of CNCC1 at 236 days at 2θ of $29.4°$ compared to the same at 56 days and cement paste is another indication of pozzolanic and hydration reaction [6]. Shaikh et al. [7] reported in their XRD results that after 28 days the cement paste containing 2% nano-silica exhibited less calcium hydroxide peaks than the control cement paste.

In nanocomposites containing 3 wt% CNC, the intensities of major peaks of Ca $(OH)_2$ were slightly decreased when compared to cement paste. For instance, the intensity of $Ca(OH)_2$ at 2θ of $18.01°$ for CNCC3 decreased by 6.9% compared to cement paste. This may be attributed to agglomerations of CNC at high contents which led to relatively poor dispersion of CNC and hence relatively poor pozzolanic reaction. This agglomeration can be explained as follows; nanoparticles, due to their small size, have high inter-particle van der Waal's forces causing them to lose the desirable specific surface area to volume ratio [2]. Therefore, due to their higher van der Waal's forces, the 3 wt% CNC agglomerate more than other 1 wt% CNC and 2 wt% CNC. As a result of this, the efficiency of 3 wt% CNC (or nanoparticles) in consuming $Ca(OH)_2$ could be less due to the reduction of total surface areas that contribute in pozzolanic reaction. Shaikh et al. [7] also stated that nanoparticles agglomerate more than other micro-pozzolanic materials (e.g. silica fume, metakaolin) due to their higher van der Waal's forces. In summary, it is important to note that the reduction of the porosity could be attributed to two reasons: (i) filling effect of CNC due to its good dispersion, (ii) the pozzolanic reaction by amorphous nanoparticles (i.e. CNC) that lead more C-S-H gel being produced.

6.2.2 Hemp Fabric-Reinforced Nanocomposites

The effect of wetting and drying cycles on the porosity of 6THFRC composites and 6THFR-CNCC nanocomposites was also shown in Table 6.1. Generally the porosity increased significantly after 3 wetting and drying cycles (at 236 days). The porosity of 6THFRC composites increased from 32.1 to 36.3% by about 13% increase. This was due to increase in voids in between fibres and matrix. Whereas in 6THFR-CNCC1 nanocomposite, the porosity increased by 8.5% compared to CNCC1 nanocomposite at 56 days. This indicates that CNC relatively reduced the number of the voids between hemp fibre and nanomatrix after 3 wet/dry cycles and thus fibre-matrix interfacial bounding was slightly maintained. Roma et al. [8] reported that in the period from 28 to 155 days, a tendency of increase in the permeable void volume was observed for the sisal fibres reinforced cement composite containing silica fume. It could be a consequence of the fast degradation of

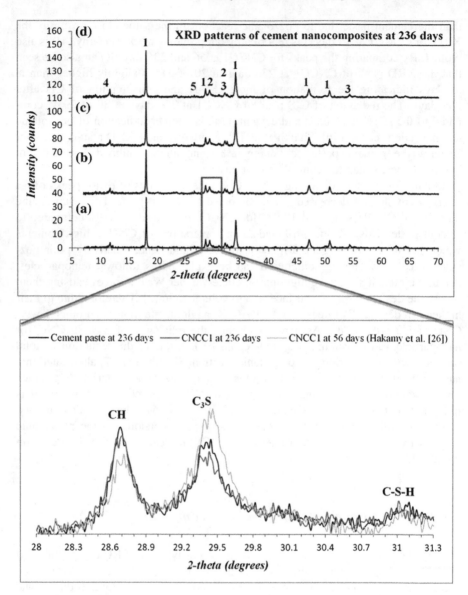

Fig. 6.1 XRD patterns of: **a** cement paste, nanocomposites containing various calcined nanoclay: **b** 1 wt% (CNCC1), **c** 2 wt% (CNCC2), and **d** 3 wt% (CNCC3). *Legend 1* Portlandite [Ca(OH)$_2$], *2* Tricalcium silicate [C$_3$S], *3* Dicalcium silicate [C$_2$S], *4* Gypsum, *5* Quartz [4]

sisal fibres in the alkaline environment of Portland cement or even due to the detachment of the cellulose fibre during wet–dry cycles attributed to its shrinkage into the cement matrix.

6.2.3 Flexural Strength of Nanocomposites

The effect of 3 wet/dry cycles on the flexural strength of cement paste and nanocomposites (CNC) is shown in Fig. 6.2. Overall, the incorporation of CNC into the cement composite led to significant enhancement in the flexural strength at all ages. At 56 days, the flexural strength of cement nanocomposite containing 1, 2 and 3 wt% CNC was increased by 42.9, 34.8 and 30.6%, respectively compared to cement paste. This improvement clearly indicates the effectiveness of CNC in consuming calcium hydroxide (CH), supporting pozzolanic reaction and filling the micro pores in the matrix [9, 10].

Thus the microstructure of cement nanocomposite is denser than the cement matrix, especially in the case of using 1 wt% CNC, which is evident from its higher flexural strength. However, after 3 wet/dry cycles at 236 days, the flexural strength of nanocomposites increased slightly compared to their values at 56 days. For example, the flexural strength of CNCC1 nanocomposite increased from 7.75 to 8.27 MPa by about 7% increase. Rong et al. [11] studied the effects of 3% nano-SiO_2 particles on the durability of concrete containing 35% fly ash at 28 and 90 days, they reported about 11% improvement in flexural strength at 90 days compared to 28 days. Mohamed [12] also reported that flexural strength of concrete containing 1% nano-SiO_2 improved from 12.08 to 13.27 by 10% after 90 days compared to its strength at 28 days.

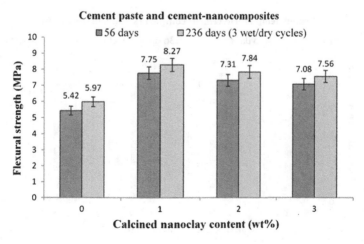

Fig. 6.2 Flexural strength as a function of calcined nanoclay content for cement paste and nanocomposites at 56 and 236 days [4]

6.2.4 Flexural Strength of Hemp Fabric-Reinforced Nanocomposites

The effect of wetting and drying cycles on the flexural strength of 6THFRC composites and treated hemp fabric reinforced-nanocomposites (6THFR-CNCC) are shown in Fig. 6.3. Generally all composite showed reduction in the flexural strength after 3 wetting and drying cycles (at 236 days). This was attributed to some partial degradation of hemp fibre in cement matrix that led to slightly deteriorate fibre-matrix bonding and also the mineralization of the fibres in which fibre became more brittle. The incorporation of CNC into the 6UHFRC composite led to significant enhancement in the flexural strength of all treated Hemp fabric reinforced nanocomposites. At 56 days, the flexural strength of 6THFR-CNCC1 increased from 14.52 to 20.16 MPa, about 38.8% increase compared to 6UHFRC composite. However, after 3 wet/dry cycles (at 236 days) the flexural strength of 6UHFRC composite dropped to 26.2% of the initial strength at 56 days, whereas the flexural strength of 6THFR-CNCC1 nanocomposites reduced by about 17.4% compered to its value at 56 days. Moreover, the flexural strength of 6THFR-CNCC2 and 6THFR-CNCC3 nanocomposites reduced by about 18.3 and 19.7% compared to their values at 56 days.

Based on this result, it can be concluded that the reduction in the flexural strength for 6THFR-CNCC3 nanocomposites was less than the reduction of 6UHFRC composite after 3 wet/dry cycles (at 236 days). This improvement is explained as follows: the degradation of natural fibres in Portland cement matrix is due to the high alkaline environment (calcium hydroxide solution) which dissolves the lignin and hemicellulose parts, thus weakening the fibre structure [13–15]. In order to improve the durability of natural fibre in cement paste, the matrix could be modified in which calcium-hydroxide (CH) must be mostly consumed or reduced.

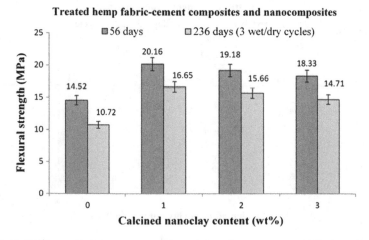

Fig. 6.3 Flexural strength as a function of calcined nanoclay content for 6THFRC composite and 6THFR-CNCC nanocomposites at 56 and 236 days [4]

In this study, the CNC effectively prevented the hemp fabric degradation by reducing the CH in the matrix through pozzolanic reaction. Thus, the degradation of hemp fibres in nanocomposite was mostly prevented and the treated hemp fabric-nanocomposite matrix interfacial bonding was mostly improved especially in the case of 1 wt% CNC. Other two reasons of this improvement were the effectiveness of CNC in filling the micro pores in the matrix which led to denser microstructure of nanocomposite matrix than the cement matrix and also the pre-treatment of hemp fibre by NaOH solution. Aly et al. [16] studied the durability of (FRCGN) flax fibre reinforced cement mortar containing 2.5 wt% nanoclay and 20 wt% ground waste glass powder at 28 days and after 50 wet/dry cycles (378 days). They reported that the FRCGN nanocomposites after accelerating ageing cycles showed small reduction (19%) in the flexural strength compared to its initial strength at 28 days. Filho et al. [17] investigated the durability of sisal fibre-reinforced mortar with 50% metakaolin (PC-MK) and without metakaolin (PC) at 28 days and after 25 wet/dry cycles. They observed that the flexural strength of PC and PC-MK composites decreased by 63% and 23%, respectively compared to their control composites at 28 days. They reported that 50% metakaolin replacement significantly prevented the sisal fibres from the degradation in cement matrix. However, the addition of CNC more than 1 wt% caused a marked decrease in flexural strength. This could be attributed to the relatively poor dispersion and agglomerations of the CNC in the cement matrix at higher CNC contents, which created weak zones and then increased the porosity.

Figure 6.4 shows the effect of wetting and drying cycles on the stress-midspan deflection behaviour of 6THFRC composites and 6THFR-CNCC1 nanocomposites. The ductile behaviour can be observed in both treated composites with and without CNC, with higher flexural strength (about 40% increase) and better post-peak ductility in the composite containing CNC. It was observed that ductile behavior and bending stress are slightly reduced due to accelerated aging. This reduction was attributed to the lignin and hemicellulose deterioration of hemp fibre in matrix by Ca^2 ions attack and embrittlement (brittleness) of hemp fibres due to fibre cell wall mineralization in cement matrix [13–15]. Moreover, alternating wet and dry conditioning increased the porosity and interface fatigue that led to decrease the bond strength between the fibres and the matrix, and thus the load-transfer. However, the 6THFR-CNCC1 nanocomposite (Fig. 6.4b) presented better ductile behavior and bending stress than 6THFRC composites (Fig. 6.4a) after 236 days. This enhancement could be due to high fibre-matrix interfacial bonding, which increases the maximum load-transfer capacity.

6.2.5 Microstructural Analysis

Figure 6.5a–d show the fibre-matrix interfacial bonding of 6THFRC composite and 6THFR-CNCC1 nanocomposite at 56 days and after 3 wet/dry cycles (at 236 days). At 56 days, the microstructural characteristic of the fracture surface of

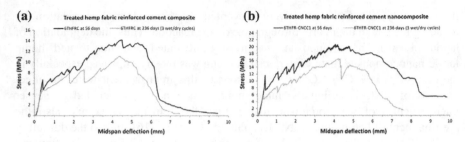

Fig. 6.4 Stress versus mid-span deflection curves for 6THFRC composite and 6THFR-CNCC1 nanocomposites at (**a**) 56 and (**b**) after 3 wet/dry cycles (236 days) [4]

6UHFRC composite (Fig. 6.5a) showed relatively good fibre-matrix interfaces with small gaps between the fabric layers and the cement matrix. However, after 3 wet/dry cycles (at 236 days) the interface zone between the fabrics and the matrix increased (Fig. 6.5b), which indicated obvious deterioration of fibre-matrix interfacial bonding. Regarding the 6THFR-CNCC1 nanocomposite at 56 days, the SEM image showed good fibre-nanomatrix interfacial bonding (Fig. 6.5c) compared to the 6THFRC composite (Fig. 6.5a). However, after 3 wet/dry cycles (at 236 days)

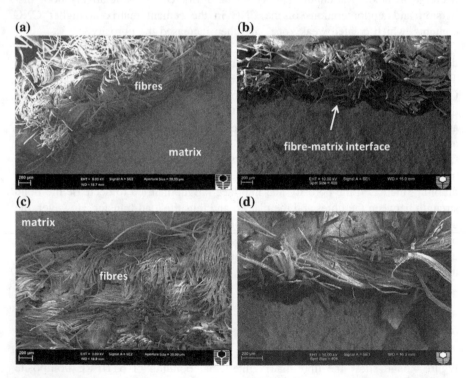

Fig. 6.5 SEM images of 6THFRC composite (**a**) at 56 days, (**b**) after 3 wet/dry cycles (236 days) and 6THFR-CNCC1 nanocomposites (**c**) at 56 days, (**d**) after 3 wet/dry cycles (236 days) [4]

the interface zone between the fabrics and the matrix containing CNC slightly increased (Fig. 6.5d), which indicated very small signs of degradation of fibre-matrix interface in nano-matrix. This confirmed that the CNC considerably reduced the deterioration of fibre-matrix interfacial bonding. The same phenomena was observed by Wei and Meyer [18], according to their SEM images where they reported that the use of metakaolin in cement composite reduced the deterioration of interface zone between natural fibre and matrix after accelerating ageing (30 wet/drying cycles).

SEM micrographs of NaOH treated hemp fibre and the hemp fibres extracted from 6THFRC composite and 6THFR-CNCC1 nanocomposite after 3 wet/dry cycles (at 236 days) are shown in Fig. 6.6(a–d). It can be seen that the fibre surface of raw NaOH treated hemp fibre was more uniform (Fig. 6.6a), whereas in 6THFRC composite (Fig. 6.6b–c), the micrographs revealed some degradation on the fibre surface during accelerating ageing. Sedan et al. [19] reported that the CH content (Ca^{2+} ions) of cement matrix contributed to the degradation of hemp fibre rapidly. Regarding 6THFR-CNCC1 nanocomposite (Fig. 6.6d), there was slight change in the fibre surface after 236 days, which indicated that CNC slightly prevented the degradation of hemp fibre surface.

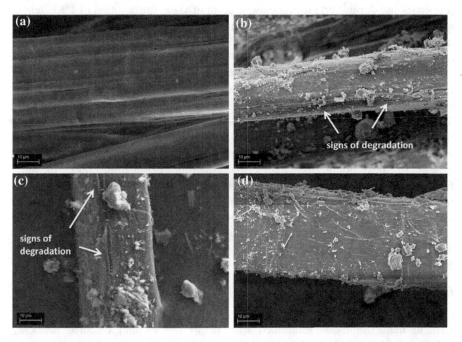

Fig. 6.6 SEM images of (**a**) NaOH treated hemp fibre and the hemp fibres extracted from: (**b**, **c**) 6THFRC composite, (**d**) 6THFR-CNCC1 nanocomposite after 236 days [4]

6.3 Cost and Applications

Natural fibres/fabrics are increasingly being utilized due to low density, low cost, renewability, recyclability and availability [13, 14]. On the other hand, nanoparticles are expensive which could limit their applications [20, 21]. Singh [22] stated that the nanomaterials have many unique characteristics which will definitely result in high strength and durable concrete, but the cost of nanomaterial is still very expensive due to novelty technology. Perhaps the nanomaterial cost will come down over time as manufacturing technologies upgrade their production efficiency [23]. However, nanomaterials are used in very small amounts in the concrete or other cementitious composites. Shaikh and Supit [24] stated that although the use of nano-$CaCO_3$ was first considered as filler to partially replace cement, some studies have shown advantages of using 1% nano-$CaCO_3$ nanoparticles in terms of compressive strength and economic benefits as compared to cement and other supplementary cementitious materials.

Among all types of nanomaterials that can be used in construction and building materials, nanoclay is considered as one of inexpensive nanomaterials. In this study the use of (very small amounts) only 1% calcined nanoclay in CNCC1 and 6THFR-CNCC1 nanocomposite led to significant improvement in mechanical properties and durability. From economic point of view, the addition of 1% calcined nanoclay in cement paste or treated hemp fabric reinforced cement composite will not add any significant cost but improved the durability significantly after 236 days (3 wet/dry cycles).

Regarding the issue of durability/cost ratio, this study has found that (see Fig. 6.5), after 236 days (3 wet/dry cycles), the 6THFR-CNCC1 nanocomposite showed higher flexural strength (by 55% increase) than control 6THFR composite. This means that the addition of 1% calcined nanoclay will not add significant cost but can increase the life (durability) of treated hemp fabric reinforced cement composites by 1.55 times over their counterparts without calcined nanoclay. The benefit of developing the long-term durability of natural fibre in such fibre-nanocomposite will overcome the few cost. The developed treated hemp fabric reinforced nanocomposites can also be widely employed as an alternative to synthetic fibres in some applications including concrete tiles, roofing sheets, sandwich panels, on-ground floors, ceilings and structural laminate.

6.4 Conclusions

Cement nanocomposites reinforced with hemp fabrics and calcined nanoclay (CNC) have been fabricated and investigated. The treated hemp fabric-reinforced cement composites and nanocomposites were subjected to 3 wetting and drying cycles and then tested at 56 and 236 days. The influences of CNC dispersion on the durability of these composites have been characterized in terms of porosity, flexural strength, stress-midspan deflection curves and microstructural observation of hemp surface. The microstructure of matrix was investigated using X-ray Diffraction. Results indicated that the CNC effectively mitigated the degradation of hemp fibres. The durability and the degradation resistance of hemp fibre enhanced due to the addition of CNC into the cement matrix and the optimum content of CNC was 1 wt%. The treated hemp fabric-reinforced nanocomposites containing 1 wt% CNC exhibited superior durability than their counterparts and exhibited good fibre-matrix interface. This environmental friendly nanocomposite can be used for various construction applications such as ceilings and roofs.

References

1. Stefanidou M, Papayianni I. Influence of nano-SiO$_2$ on the Portland cement pastes. Compos Part B. 2012;43:2706–10.
2. Supit SWM, Shaikh FUA. Durability properties of high volume fly ash concrete containing nano-silica. Mater Struct. 2014;47:1545–59.
3. Jo B, Kim C, Tae G, Park J. Characteristics of cement mortar with nano-SiO$_2$ particles. Constr Build Mater. 2007;21:1351–5.
4. Hakamy A. Microstructural design of high-performance natural fibre nanoclay cement nanocomposites. Ph.D. Thesis Curtin University, Perth Australia 2016.
5. Hoshino S, Yamada K, Hirao H. XRD/Rietveld analysis of the hydration and strength development of slag and limestone blended cement. J Adv Concr Technol. 2006;4:357–67.
6. Wei Y, Yao W, Xing X, Wu M. Quantitative evaluation of hydrated cement modified by silica fume using QXRD, 27Al MAS NMR, TG–DSC and selective dissolution techniques. Constr Build Mater. 2012;36:925–32.
7. Shaikh FUA, Supit SWM, Sarker PK. A study on the effect of nano silica on compressive strength of high volume fly ash mortars and concretes. Mater Des. 2014;60:433–42.
8. Roma L, Martello L, Savastano H. Evaluation of mechanical, physical and thermal performance of cement-based tiles reinforced with vegetable fibres. Constr Build Mater. 2008;22:668–74.
9. Shebl S, Allie L, Morsy M, Aglan H. Mechanical behavior of activated nano silicate filled cement binders. J Mater Sci. 2009;44:1600–6.
10. Li H, Xiao H, Yuan J, Ou J. Microstructure of cement mortar with nano-particles. Compos Part B. 2004;35:185–9.
11. Rong Z, Sun W, Xiao H, Jiang G. Effects of nano-SiO$_2$ particles on the mechanical and microstructural properties of ultra-high performance cementitious composites. Cem Concr Compos. 2015;56:25–31.
12. Mohamed A. Influence of nano materials on flexural behavior and compressive strength of concrete. HBRC Journal. 2015. doi:10.1016/j.hbrcj.2014.11.006.
13. Ardanuy M, Claramunt J, Filho R. Cellulosic fibre reinforced cement-based composites: a review of recent research. Constr Build Mater. 2015;79:115–28.
14. Pacheco-Torgal F, Jalali S. Cementitious building materials reinforced with vegetable fibres: a review. Constr Build Mater. 2011;25:575–81.
15. Odler I, Wonnemann R. Effect of alkalies on Portland cement hydration: I. Alkali oxides incorporated into the crystalline lattice of clinker minerals. Cem Concr Res. 1983;13:477–82.
16. Aly M, Hashmi M, Olabi A, Messeiry M, Hussain A, Abadir E. Effect of nano-clay and waste glass powder on the properties of flax fibre reinforced mortar. J. Eng. Appl. Sci. 2011;6:19–28.
17. Filho J, Silva F, Filho R. Degradation kinetics and aging mechanisms on sisal fibre cement composite systems. Cem Concr Compos. 2013;40:30–9.

18. Wei J, Meyer C. Degradation mechanisms of natural fibre in the matrix of cement composites. Cem Concr Res. 2015;73:1–16.
19. Sedan D, Pagnoux C, Smith A, Chotard T. Mechanical properties of hemp fibre reinforced cement: influence of the fibre/matrix interaction. J Eur Ceram Soc. 2008;28:183–92.
20. Pacheco-Torgal F, Jalali S. Nanotechnology: advantages and drawbacks in the field of construction and building materials. Constr Build Mater. 2011;25:582–90.
21. Hakamy A, Shaikh FUA, Low IM. Effect of calcined nanoclay on the durability of NaOH treated hemp fabric reinforced cement nanocomposites. Mater Des. 2016;92:656–66.
22. Singh T. A review of nanomaterials in civil engineering works. Int J Struct Civ Eng Res. 2014;3:31–5.
23. Hanus MJ, Harris AT. Nanotechnology innovations for the construction industry. Prog Mater Sci. 2013;58:1056–102.
24. Shaikh FUA, Supit SWM. Mechanical and durability properties of high volume fly ash (HVFA) concrete containing calcium carbonate ($CaCO_3$) nanoparticles. Constr Build Mater. 2014;70:309–21.

Chapter 7
Summary and Concluding Remarks

7.1 Nanoclay-Cement Nanocomposites

Based on the XRD and SRD results, the addition of nanoclay reduced the intensities of major peaks of $Ca(OH)_2$ crystals comparing to the control cement paste. Also from QXDA, the addition of 1 wt% nanoclay reduced the amount of $Ca(OH)_2$ from 19.5 to 15.7 wt%, about 19.5% reduction compared to cement paste. Also the intensities of major peaks of $Ca(OH)_2$ were significantly reduced compared to cement paste. Furthermore, the amorphous content was increased slightly from 67.4 to 70 wt%, about 3.7% increase. This indicates that an obvious consumption of $Ca(OH)_2$ crystals mainly due to the effect of pozzolanic reaction in the presence of nanoclay and good dispersion of nanoclay in the matrix leads to produce more amorphous calcium silicate hydrate gel (C-S-H). In addition, SEM images of the microstructure of cement paste, nanocomposites containing 1–3 wt% nanoclay were also presented. SEM images of cement paste showed more $Ca(OH)_2$ crystals and Ettringite as well as more pores which revealed a weak structure. It was found that the SEM micrograph of nanocomposites containing 1 wt% nanoclay was different from that of cement paste, the structure was denser and compact with few pores due to the filler effect of 1 wt% nanoclay and more CSH gel due to pozzolanic effect. On the other hand, the SEM micrograph of nanocomposites containing 3 wt% nanoclay showed more pores and micro-cracks which led to weaken the structure of this nanocomposite when compared to nanocomposites containing 1 wt% nanoclay.

The incorporation of nanoclay in cement matrix had led to a modest enhancement in physical and mechanical properties of all nanocomposites. Moreover, it was found that the nanocomposites with 1 wt% of nanoclay achieved the highest improvement in all properties. The cement nanocomposite containing 1 wt% nanoclay decreased the porosity (by 20.6%) and water absorption (by 23.5%) and increased the density (by 4%), compressive strength (by 31%), flexural strength (by 32%), fracture toughness (by 31%), impact strength (by 29%) and Rockwell

© The Author(s) 2017
I.-M. Low et al., *High Performance Natural Fiber-Nanoclay Reinforced Cement Nanocomposites*, Biobased Polymers, DOI 10.1007/978-3-319-56588-0_7

hardness (by 24%) as well as improved thermal stability (by 2.3%) compared to the control cement paste. This improvement can be attributed to pozzolanic and filler effects of 1 wt% nanoclay in which this nanocomposite had more consolidated microstructure than others. However, the addition of more nanoclay into cement nanocomposite adversely affected the mechanical and thermal properties. This could be attributed to poor dispersion and agglomerations of the high nanoclay contents which created more voids in the matrix.

7.2 Hemp Fabric-Reinforced Nanoclay-Cement Nanocomposites

The addition of nanoclay into hemp fabric-reinforced nanocomposite improved the physical, mechanical and thermal properties. Moreover, the hemp fabric-reinforced nanocomposites containing 1 wt% nanoclay achieved the highest improvements than others. It was found that the incorporation of 1 wt% nanoclay into the hemp fibre-reinforced nanocomposites decreased the porosity (by 16%) and water absorption (by 18%) as well as increased the density (by 3%), flexural strength (by 28.5%), fracture toughness (by 24.6%), impact strength (by 23%) and also improved the thermal stability (by 19%), when compared to the control hemp fibre-reinforced cement composite.

SEM micrographs of fracture surfaces of hemp fibre-reinforced nanocomposites containing 1 wt% nanoclay showed good fibre-matrix interface as well as the presence of hydration products on the fibre surface indicating better fibre/matrix interface bond. However, poor adhesion between fibres and matrix was observed in hemp fibre reinforced cement composite. In hemp fibre reinforced-nanocomposite containing 3 wt% nanoclay, macro-cracking was observed in the matrix which revealed relatively weak matrix and also debonding of fibre occurred.

This improvement in physical and mechanical properties as well as fibre-matrix interface of hemp fibre-reinforced nanocomposites containing 1 wt% nanoclay was attributed to the fact that the nanoclay modified the microstructure of matrix through pozzolanic reaction and pore-filling effect. Thus, good interfacial bonding between the nanomatrix and the hemp fibres was achieved. However, the addition of more nanoclay (> than 1 wt%) into the hemp fibre-reinforced cement composites adversely affected the thermal, physical and mechanical properties as well as fibre-matrix interface.

7.3 Calcined Nanoclay-Cement Nanocomposites

Based on XRD, EDS and HRTEM results, nanoclay transformed to amorphous state (calcined nanoclay) at 900 °C. Moreover, at high magnification by HRTEM, it was found that many platelets in calcined nanoclay were destroyed and some of them broke to small nanoparticles with approximate spherical shapes ranging 3–8 nm. Overall, the results showed that the microstructures, physical, mechanical and thermal properties of the cement nanocomposites were improved as a result of CNC addition. The optimum content of CNC was 1 wt%.

Regarding characterisation of microstructure for nanocomposites, the results of XRD and QXDA indicated that the addition of 1 wt% CNC reduced the amount of $Ca(OH)_2$ from 16.8 to 12.1 wt%, about 28% reduction compared to cement paste. Also the intensities of major peaks of $Ca(OH)_2$ were significantly reduced compared to cement paste. Furthermore, the amorphous content of C-S-H gel was increased from 70.1 to 74.8 wt%, about 6.7% increase. This improvement in microstructure can be explained as follows: the CNC was mainly amorphous nanomaterial and behaved as a highly reactive artificial pozzolan and thus it could consume more CH through the pozzolanic reaction. As a result the CH content in cement nanocomposite containing 1 wt% CNC was reduced significantly when compared to cement paste and other cement nanocomposites containing NC and CNC such as cement nanocomposite containing 1 wt% NC. SEM examinations of the microstructure of cement paste, 1 wt% CNC nanocomposites and 3 wt% CNC nanocomposites confirmed the effectiveness of 1 wt% CNC. SEM micrograph of 1 wt% CNC nanocomposite showed that its microstructure was denser and more compact with fewer pores, less CH and more C-S-H gel when compared to cement paste and 3 wt% CNC nanocomposites.

Regarding the physical, mechanical and thermal properties of cement nanocomposites, overall, the addition of CNC into the cement matrix led to significant enhancement in these properties when compared to cement nanocomposites containing NC or cement paste. The cement nanocomposite containing 1 wt% CNC decreased the porosity (by 31.2%) and water absorption (by 34%) and increased the density (by 9.7%), compressive strength (by 40%) flexural strength (by 42.9%), fracture toughness (by 40%), impact strength (by 33.6%) and Rockwell hardness (by 31.1%) as well as improved thermal stability (by 3.3%) compared to the control cement paste. This improvement was due to the effective filler and pozzolanic reaction effects and good dispersion of 1 wt% CNC. In fact, it could be recommended that more research is needed to overcome the agglomerations of NC or CNC and to identify the best method of mixing to achieve good dispersion of CNC in the matrix.

7.4 Un-Treated and NaOH Treated Hemp Fabric-Reinforced Cement Composites

Based on SEM, XRD and TGA results of hemp fibre surface, it was found that NaOH treatment effectively bleached and cleaned the surface of hemp fibres and removed the amorphous materials such as hemicellulose, pectins and impurities (fatty substances and waxes) from their surface. Thus the fabric surface became more uniform with slightly higher crystallinity index and thermal stability when compared to un-treated hemp fabric surface.

Regarding the flexural strength and fracture toughness of un-treated hemp fabric-reinforced cement composites (UHFRC), both flexural strength and fracture toughness increased with an increase in fibre content up to the optimum hemp fibre content, and then decreased after this limit. This suggests that the optimum hemp fabric content among these composites was 6.9 wt% for 6UHFRC composite. In 6UHFRC composite, the flexural strength and fracture toughness were improved by 133–303% respectively compared to the cement paste. However, beyond this optimum content of hemp fibre, both flexural strength and fracture toughness of the UHFRC composites were decreased. This could be attributed to an increase in porosity which reduced adhesion between the fibres and the matrix, and thus decreased the load transfer.

NaOH treatment of hemp fabric had a positive effect on the flexural strength and fracture toughness of 6THFRC composites. Flexural strength of 6THFRC composite increased from 12.6 to 14.5 MPa, about 14.9% increase when compared to 6UHFRC composite. Moreover, the 6THFRC composite exhibited 13.5% increase in fracture toughness.

SEM results of 6UHFRC composite showed relatively poor fibre-matrix interfaces with small gaps between the fabric layers and the cement matrix. In contrast, the 6THFRC composite showed very small gaps between the fabrics and the matrix, which indicated better fibre-matrix interfacial bonding. This improvement could be explained as follows: after NaOH treatment, most of waxes and fats were removed from the hemp fibre surface, and the surface became more uniform but rough. Thus, this led to good interfacial bonding between the matrix and the hemp fibres which served to enhance the load transfer process at the interface.

7.5 NaOH Treated Hemp Fabric Reinforced Calcined Nanoclay-Cement Nanocomposites

Physical, mechanical and thermal properties were enhanced due to the addition of CNC into the 6THFRC composites and the optimum content of CNC was 1 wt%. The treated hemp fabric-reinforced nanocomposites containing 1 wt% CNC (6THFR-CNCC1) exhibited the highest flexural strength, fracture toughness, impact strength and thermal stability than their counterparts and good fibre-matrix interface. For 6THFR-CNCC1 nanocomposites, it was found that porosity and water absorption decreased by 12.4–14% respectively. Values of density, flexural strength, fracture toughness, impact strength and thermal stability improved by 4.5, 38.8, 38.1, 19 and 5% respectively. 6THFR-CNCC1 nanocomposite showed better fibre-nanomatrix interfacial bonding with ruptured fibres when compared to 6THFRC and 6THFR-CNCC3 composites.

This improvement clearly indicated the effectiveness of CNC in consuming calcium hydroxide (CH), thus facilitating the pozzolanic reaction, and pore-filling in the matrix. As a result, the microstructure of nanocomposite matrix was denser than the cement matrix. Consequently, the interfacial bonding of treated hemp fabric-nanocomposite matrix was mostly improved, especially in the case of using 1 wt% CNC, as evident from its higher strength and toughness values. It could be seen that CNC behaves not only as a filler to improve the microstructure, but also as the activator to facilitate the pozzolanic reaction and thus improved the adhesion between the treated hemp fabric and the matrix. However, the mechanical properties of treated hemp fabric-reinforced composites were adversely affected when CNC content increased beyond the optimum content of 1 wt%. This was attributed to the poor dispersion of high content of CNC into the matrix, which led to increase in porosity and weaken the fibre-matrix interface.

7.6 Durability of NaOH Treated Hemp Fabric Reinforced Calcined Nanoclay-Cement Nanocomposites Subjected to Wet/Dry Cycles

Regarding the durability of calcined nanoclay-cement nanocomposites, porosity and flexural strength improved slightly between 56 and 236 days where the flexural strength of CNCC1 nanocomposite increased by 7%. This improvement could be attributed to further hydration reaction and pozzolanic reaction ands pore-filling effect. In NaOH treated hemp fabric reinforced calcined nanoclay-cement nanocomposites, the durability and the degradation resistance of hemp fibre was enhanced due to the addition of CNC into the cement matrix and the optimum content of CNC was 1 wt%. After 236 days, the flexural strength of 6THFRC composites decreased by 26.2% compared to its initial strength at 56 days. In comparison the flexural strength of 6THFR-CNCC1 nanocomposites decreased by only 17.4%.

SEM results after 236 days indicated that there were very small signs of degradation of fibre-matrix interface for 6THFR-CNCC1 nanocomposites when compared to 6THFRC composites. Moreover, the SEM images of hemp fibres extracted from 6THFRC composites showed more degradation than hemp fibres extracted from 6THFR-CNCC1 nanocomposites.

Overall, the addition of CNC had great potential to improve the durability of treated hemp fabric reinforced cement nanocomposites during wet/dry cycles, particularly at 1 wt% CNC. In order to improve the durability of natural fibre in cement paste, the matrix could be modified in which calcium-hydroxide (CH) must be mostly consumed or reduced. In this study, the CNC effectively prevented the hemp fabric degradation by reducing the CH in the matrix through pozzolanic reaction. Thus, the degradation of hemp fibres in nanocomposite was mostly prevented and the treated hemp fabric-nanocomposite matrix interfacial bonding was mostly improved especially for 1 wt% CNC.

7.7 Concluding Remarks

High performance natural fibre-reinforced cement nanocomposites have been suc-
cessfully fabricated. The effect of nanoparticles (nanoclay and calcined nanoclay)
and hemp fabrics (untreated and NaOH treated) on the microstructure, physical,
mechanical and thermal properties as well as durability of cement nanocomposites,
untreated and NaOH treated hemp fabric-reinforced cement nanocomposites were
investigated and discussed. The research results provided a fundamental knowledge
on the mechanism and performance of a new class of treated hemp fabric-reinforced
cement nanocomposites. Despite the significant improvement in mechanical properties for both cement nanocomposites and NaOH treated hemp fabric-reinforced
cement nanocomposites, the agglomeration of nanoparticles at high content is still a
major issue. In addition, advanced computer models are required to investigate the
influence of nanostructures, such as shape and size distribution, orientation, aspect
ratio, degree of spatial and interfaces on the physical and mechanical properties of
cement nanocomposites, untreated and treated hemp fabric-reinforced cement
nanocomposites. Multi-scale mechanics models and numerical methods should be
developed for better understanding of the enhanced properties in these building
nanomaterials.

Printed in the United States
By Bookmasters